OCTAVIO PEREZ-BEATO PH.D.

Genetic Variation, Evolution, and Creation

the unfolded truth

ISBN: 1-4611-7348-5
ISBN-13: 9781461173489

DEDICATION

To the loving memory of my wife, Aracely. For more than forty-eight years she was my support in every way. Her life was an example of courage, altruism, and love.

CONTENTS

INTRODUCTION

Many questions may arise regarding the beginning of times, the universe, planet Earth, and living things on it, including the granted concept of humanity as the superb ending result of evolution. Many attempts, thousands perhaps have been done through science and religion to understand...I mean, *try to understand* what we are, what the surrounding living world...or worlds are, and how everything began in the darkness of time. Hundreds of questions may arise to exactly realize where we are now, and what the biological, social, and cultural future shall be.

The literary and scientific boom of the nineteenth century *On the Origin of Species* has not convinced, as a whole, nowadays-modern scientists whatsoever. Unfortunately, Charles Darwin failed to consequently achieve what the title of his then masterpiece announced to prove. He only, and not extensively, showed the intricate and incomplete path of species evolution, and not "the origin of species."

Understandable efforts with results, at the beginning of the twentieth century, spilled doubts on the table about the strict mechanism of evolution, as stated in the aforementioned Darwin's exceptional book. Among others, paleontology finds increasing quality day by day, failed to find a genuine and logical systematic, gradually consistent evolution on studied species in the layers of rocks all over the world, as it would happen according to *gradual evolution* stated by Darwin. Such an inconsistency launched a profound shadow of doubt on the apparently simple explanations contented in On the Origin of Species.

A revision in full details up to great lengths was mandatory, according to tidy and thoughtful, non-fanatic scientists. The knowledge on genetics was increasing dramatically after the rediscovery of Mendel's research. Besides, mathematical formulation and approaches were taking place tightly linked to the newborn science of Genetics. Very soon on the horizon appeared the first sails of Population Genetics, which added a new dimension to the problem of evolution. A new approach was at reach, the *synthesized theory*, Evolution and Genetics, both together to enlarge the incomplete theoretical landscape of species evolution. The first incognita, "the origin of species" remained unsolved. At the same time, more questions arose on the way.

On the other hand, religion made extraordinary efforts to discredit evolution theory as opposed to orthodox belief systems; an attempt firmly related to the discussion of our own essence. The issue reached such an absurd point as to deny...or better, ignored recent and undisputable anthropological discoveries as those achieved in Sterkfontein, Africa, which shed a light on the evolution of human race. The shortsighted people stuck to the idea that evolution and creation were opposite points of views. This kind of blindness forbade the simplest reasoning on the matter. What was the reason to interpret evolution and creation as opposite concepts? Was evolution a menace to the established and orthodox belief systems? How could those people overlook the impressive finds of Australopithecine skeleton remains resembling those of humans in many ways? Were these forerunners of *Homo sapiens* a real threat to belief systems?
Science creates thousands of books in a prolific way exposing an open controversy; most belief systems based everything in just one single ancient book: the bible.

The science is unstoppable, and new theories jumped into light based on new discoveries. The most advanced scientists projected theoretical proposals to explain the "missing links" in evolution. But it is worth to say that missing links were not confined only to human evolution. Missing links are the order of modern paleontology, a real proof of non-gradual evolution that instead seems to appoint that organisms reach higher levels of development "suddenly" in time. Humans are apart from chimpanzees in less than 2 percent of genetic differences. What a marvelous difference from ape to man! Would anybody dare to question the reality of such affirmation? This is not a matter of believing, it is a scientific fact, chimp and human chromosomes and genes have been studied on a comparative basis for many years. Any mid-educated person knows it is a scientific truth. Then again, why some others denied such a fact? We are unquestionably related.

Evolution is not as many people think it is. It is not a matter of a systematic slow pacing process. Genetic variations in the same animal populations and the emergence of species in many cases prove otherwise. Besides, is our genetic endowment really completed? Are we perfect beings from a genetic point of view? Molecular genetics techniques are really advanced today, and science is facing other matters and questions very different from those science faced fifty years ago. For modern scientists, it is not only a matter of how to solve certain unknown problems on the essence of humankind. For them, it is also important to determine what pitfalls and flaws are or has been present in the evolutionary process. All this effort is aimed to explain our most controversial behavioral patterns, individually, as a whole population, and as a species as well.

It is incredible how many erratic pathways, trial and error, undetermined courses, and inexplicable genetic structures have been found as undeniable facts in present evolutionary studies, mainly in molecular genetics. All these have proved that many aspects of human life on our planet ought to evolve in someway; we are not a completed task. To grant humankind the "deserved title" of **Supreme Being on Earth** requires, at present, a completion in our genetic, cultural, and social structural plan. We are far apart from that position. Our biosocial status does not match with our behavioral pattern. It seems like some hidden inherited genes are present in our genetic background to disturb the harmonic relation that must exist on the human triad: *biology-social being-behavior.* We are not aware that we are a population on the planet, one and only one population. Unfortunately, there are many examples around the world of people believing that is more important to be a member of a certain belief system, or citizen of a particular country, or be born in the realm of a selected caste than to accept ourselves as a unique population on planet Earth, our home planet; any other considerations are irrelevant. This reality is such that albeit we are able to express ideas and emotions, and create highly advanced technologies, an inexplicable threshold behavior pushes us toward our hominoid forerunners behavioral manners. All these are threatening our own existence and evolution as human beings, for we act as merely evolved animals in many ways.

Nowadays, institutions and scientists are involved not only in the discovery of our biological, cultural, and social history far beyond present. They are really concerned about our own genetic shortcomings, and the consequences this might account for in times to come. Can genetic variation and evolution explain our present pitiable genetic status, or is creation in some way the top explanation to really understand our role on earth as highly evolved creatures? Or is it a combination of all these?

The Author

Chapter 1
The Molecular Level as an Origin.

Manfred Eigen, the 1967 Nobel Prize in Chemistry, in his outstanding book *Stufen zum Leben*, (Steps Towards Life, 1992) openly stated that no formula could be used to really deduce the origin of life, and to explain its marvelous variety from viruses to human mind. Besides, our beloved planet, beautiful and extraordinarily full of charm, which is rooted in its stunning landscapes in so many islands, continents, rivers, lakes and lagoons. This list may continue on beautiful mountain ridges, mild sloping hills, sunsets and dawns. All this astounding beauty is seen from pole to pole, from east to west, including tropical and temperate zones, in woods and forests, and in coral reefs. It is a gift to all living creatures including humans that populate this spherical wonder.

But, our Earth is just like a granule of sand in the immense universe where an uncountable amount of solar systems, included in galaxies, gigantic ones compare to our Milky Way galaxy, do exist. Now, scientists seem to agree that the universe is about 13.7 billion years old. For me, the calculation may run out of logic, we only know a minuscule portion of that universe, though now we count on the Hubble Space Telescope, which can capture images of incredible old galaxies, as old as 13.1 billion years. It is very close to the "calculated age" of the universe, if it is at all.

As the molecular world is present everywhere in the universe, I will try to expose the ascending knowledge on molecular structures. The reader should extract his/her own conclusions at the light of critical thinking, without fanaticism in one way or another.

It is considered the turning point on Molecular Biology the discovery in 1953 by Rosalind Franklin, a physical chemist and an x-ray crystallographer specialist, of the first two forms of DNA, which she called A and B. She wrote in her laboratory records that DNA was a two chains molecule. She gently offered her relevant photograph to Watson and Crick, who in turn arrived to the double-helix DNA model, based on that unique photograph. Unfortunately, Franklin died in 1958 and the Nobel Prize was not reached by her hands; the hands that conducted the unprecedented x-ray crystallography, evidencing the two-stranded DNA molecule for the first time in science history. The truth is meant to be openly expressed; it is not to be concealed.

The DNA molecule is the hereditary material carried by all living creatures on earth, including plants of course. It is not the purpose of this book to explain in full details how the structure of this molecule is, you may refer to any book of Biology to check details you may be interested in. Schematic explanations will be used to simplify the high complexity of chemical compounds and their bonds, just to be accessible to all readers, no matter what academic profile they have.

As DNA is a double-strand chemical structure, one strand is connected to the other by chemical bonds between bases; one group of bases is called *purines* and the other *pyrimidines*. One base from the pyrimidine group in one strand bonds to a base from the purine group in the other strand, no matter if they belong to one strand or the other; they are complementary bases, so they are paired. Besides, there are about 10 base pairs for each turn of the helix structure. Each strand serves as a template to build a new complementary strand (Fig. 1).

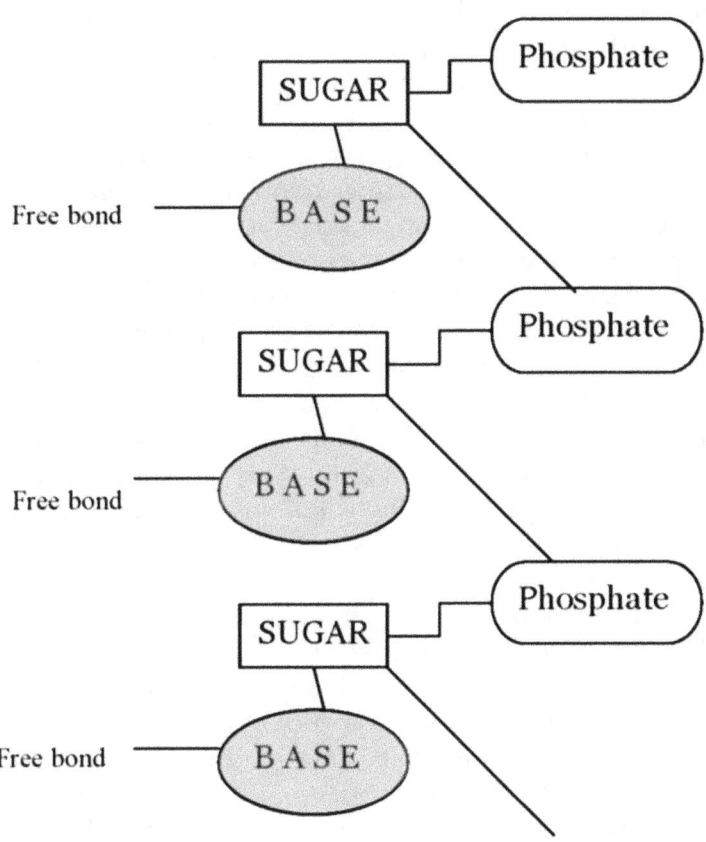

Fig. 1 Schematic sequence of three nucleotides and their sugar-phosphate chemical bonds, joining them. Free bonds ready to join the other strand.

The sugar present in DNA is known as Deoxyribose, hence the name Deoxyribonucleic Acid (DNA), and the one present in RNA is Ribose (Ribonucleic acid). It is important to explain that any molecular unit comprising a purine or pyrimidine base, a deoxyribose or ribose (both are sugar compounds), and one or more phosphate groups, is known as a *nucleotide*. A phosphate group is a molecule comprising a phosphorus atom connected to a couple of oxygen atoms and two OH- ions, which forms a salt, the phosphate. Now, the DNA strands could be huge, made of millions of nucleotides in length; actually an immense amount of messages are contained in such molecules. Although there are only four kinds of bases in the double helix, what becomes amazing is the amount of messages they can provide when bonded to the nucleotides. Let us use an illustrative case. If we consider a short sequence of 10 different nucleotides, then 4^{10} is the amount of possible different combinations or messages; more than a million in just such a short sequence. Now, if we remember, the DNA strands may contain thousands of millions of nucleotides, it is simple to figure out the huge amount of genetic messages included in such strands. This information enclosed in the sequence of nucleotides, will be translated into a specific linear sequence of protein units.

Stop just for a minute to think how simple a nucleotide is, and in its own simplicity it is gigantic for the amount of messages it is loaded with. Now, consider that such a marvelous structure, with such an imponderable efficiency, is able to carry the most inconceivable account of biological coded instructions to create a living organism, merely through a process of *trial and error*, nothing else; such a wonder came to be using exactly the available components that existed in the darkness of time. Did all these happened just by a simple chance of occurrence? In summary, there were the *ad hoc* components plus a blind process of trial and error, which originated the DNA extraordinary molecule, irrespective how much time went by, probably thousands of million years. Here, two inquisitive questions are pertinent:was the ancient environment so stable and steady in time to guarantee those extraordinary chemical reactions to create the DNA molecule if it took thousands of million years? The longer the time it took, the less probable the steadiness of conditions in the ever changing climate and environment, in our early Earth. But it is not only the structure of DNA itself what is surprising; the amazing fact is the biological function of that molecule to make life possible. Was it just a random process, with the available chemical compounds at the time that ultimately determined life existence on Earth?

We know for certain that chemical reactions need steady environmental conditions to proceed. Another possible question is: Is it also a mere coincidence that the three basic components of nucleotides were available for the processes of Natural Selection and Evolution to take advantage, and then to start assembling the remarkable efficient DNA? I think we have to be very ingenuous to reduce such a transcendental event to a simple *trial and error* process, which is ultimately what natural selection is. I am not denying the process of Natural Selection, it is a fact; what seems to be candid is just believing this is the one and only explanation to convince our intellect that the evolutionary process is a probabilistic, random selective phenomenon, with no more ingredients; too simplistic for the *grandeur* of its significance. Then again, if it was a random process, why are we so interested to find the natural laws that made possible life on Earth? Where do they come from? Some evidences point to the event that, probably, the most primitive life forms on Earth began some 700 million years after our planet was already formed. There are several theories to explain how this could happen. One of those theories is based on the so-called inter-stellar clouds transporting microbes as a primitive life form. Anyway, this and other theories are not appropriate to explain, convincingly, the origin of life. What is very clear to many scholars is that the emergence of life is not supported by the sole structure and function of a single molecule, but due to a complex of cooperative multi-molecule-property system, with full interaction; enzymes are of a great significance.

Let us go back again to DNA. Now, the process of DNA replication involves very special enzymes to break up the bonds between bases on each DNA strand, leading to the unwinding of both strands. During the process called *transcription,* one of the

DNA strands works as a template to create a molecule of messenger RNA (mRNA). This mRNA is in fact also a template to create a specific protein. It is worth to say that mRNA is complementary to its DNA template.

In a very simplified fashion, this is the way that a protein is built from the message contained in the DNA. The biochemical process is much more complex than the explanation offered here, which has been purposely reduced to make it comprehensible to non-specialist readers, yet clear enough to show the mechanism of protein construction.

At this point, I would like to add, beyond the extraordinary three-dimensional structure of the DNA, the fact that all this mechanism of protein production is intended to create different types of proteins with their 3-D structures comprising its form and function, based on the sequence of amino acids from which the protein is made of. Such a sequence was created according to the exact information on the DNA, through the process we have discussed before. The expression of such information leads to the development of the different aspects of an organism appearance. It is important to emphasize at this point, beyond the fact that it is unknown, how all this information is converted into organs, head, legs and so forth. There is a gap between the production of the protein and the whole formation of the living being. We only know that the coded information in the DNA is the genetic material, but we do not know how the elements that differentiate the eyes from the nails, work. It is not just to create proteins on a genetic basis, the core of the problem is the living organism itself, and we do not know exactly how is the mechanism to ensemble the living being with all the perfection that is accomplished in such a task, from simple genetic instructions. Furthermore, all the exact and almost perfect sequences at the embryological level are determined by the information coded in the DNA molecule. What a perfection to rely exclusively on a process of trial and error, no matter how many thousands of millions of years it took to be accomplished.

To add more elements to the aforementioned analysis, it is convenient to recall what researchers have carried out to better understand the process from an early embryo to a completed animal. Dr. Lewis of Caltech working with the fruit fly could determine that a very simple mechanism was responsible to shape the body of that insect: a single gene determined each segment of the fly. Later, Levine and McGinnis could prove that a gene they called *antennapedia* turned into a band just in the middle of the fruit fly embryo. That gene was acting like a toggle activating the segment that would become the thorax of the fly. Those were excellent results from really thoughtful scientists, but not enough to answer other questions that still remain with no response.

The problem is not exclusively that there are genes responsible to form a certain part of the body, as was demonstrated in the fruit fly; the point is how those genes align themselves to create a coordinated assemblage of the whole fly. It is just like the piano and the music paper. You have the piano keys, each one with a certain sound, no matter what the arrangement of the keys is, the sound is in them: they are the genes. What guide the player's fingers are the notes on the music paper, which are coordinated and arranged to produce the symphony; the symphony shall never be, otherwise. Besides, the functional coordination of all biological systems is what makes the organism an integral efficient unit, from the tiny unicellular protozoa to the huge dinosaurs. They are amazing bio-functional engines.

As has been stated by other specialists, the evident circular interdependency of all the aforementioned events, as the replication of nucleotides due to the presence of specific enzymes (which are proteins themselves), and the resolved of enzyme structures due to the presence of the right nucleotide sequences, is the core for the many difficulties in finding the way that may lead to a plausible explanation on the origin of life. Among several questions, some very simple may arise: How those biochemical agents arose? Which of them came first, and how long did it take to appear the others? How the first could survive in a changing world until the others showed up over millions of years?

The environmental conditions on earth during such an immense lapse had to be changing in such a way that, what was convenient for a specific chemical process during thousands of years, probably much more, was not convenient for the next necessary sequence to create such molecules and chemical compounds, to continue its evolutionary and unstoppable process. Remember that we are talking about processes that endured for thousands of million years. Imagine how many and different environmental changes could occur in such a gigantic lapse, and then the so-called adequate cultures for chemical compounds to thrive could disappear *before* the next logical step would be possible.

As a conservative figure about the age of our planet, it is around 4,800 million years. Furthermore, many specialists agree that from the first single-cell organism to humans, it took about 3,500 million years. Then again, the problem is that very specific conditions were necessary to go from molecules to unicellular organisms. Other very different conditions on earth were necessary to transit from single-cell organisms to metazoans (multicellular organisms), no matter how simple those metazoans were. From those ancient metazoans to the emergence of more complex organisms, it required other specific environmental conditions, different from the previous ones necessary for the preliminary processes, and so forth up to the giant dinosaurs.

How so many special conditions could be present at the exact moment to situate one organism on the pathway toward more evolved forms? If those extraordinary conditions were not present at the precise moment in the evolutionary process, probably those populations (proteins, single-cell organisms, archaic metazoans, etc.) would irreversible cruise into extinction before any further evolutionary steps could be accomplished; so much more if reckon that those evolutionary changes were separated hundreds of million years from one another.

Now, today science realizes that genes, as we know them, were established not by a random process, but by some optimization process. To declare this is just like saying that an optimal pathway, with very high functional parameters, would lead this specific and peerless process of genetic information, from the most elemental steps up to the uppermost work of biological art.

Chemists, almost around the world, stated that due to the changing nature of all these primordial chemical processes, stable conditions were not the rule, on the contrary, a favored selection (due to Natural Selection) on an important molecule from the information it carries, is not based on structural stability, it is rather based on the order that is inherent to the dynamics of its own reproduction as a molecule. It is worth to say that DNA is not 100% stable, as we may see in future pages. If DNA would be a stable molecule, evolution could not be possible, for the magical feature of DNA is precisely a certain grade of flexibility regarding sequences of nucleotides, positions and replication of specific genes, and the emergence of new ones: the mutation process. Fig. 2 depicts a very basic and simple diagram of the coordinated course in the production of a simple protein, as has been explained in previous paragraphs.

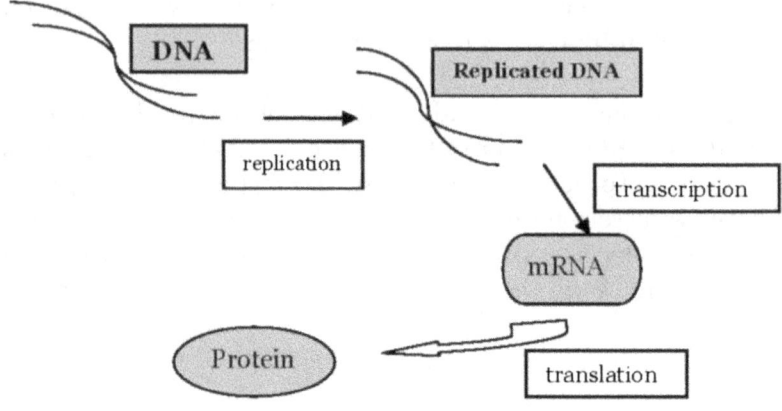

Fig. 2 Schematic representation from DNA coded message up to the formation of a protein as a final product.

Amino acids are the bricks that construct proteins, and there are 20 amino acids present in nowadays proteins. The probability of placing the best amino acid in the right position in a protein chain, without the intervention of selection is 5%, then 95% will be placed by selection, in such a way that each position will be occupied by the correct amino acid, allowing selection to occur to place the next amino acid in the right position as well. Finally, it is not the same to randomly place the amino acids all at once to form the chain, than to place one by one by selection. Every time a successful amino acid is placed in the correct position, the probability of positive selection increases for the whole chain. What a powerful mechanism that of Natural Selection! If it is at all, how long will it take for a DNA chain to be successfully constructed by the above mechanism of placing the whole sequence of amino acids with the approval of Natural Selection? Beyond that, the conditions for Natural Selection to operate are self-reproduction, mutagenicity, and an efficient metabolism; according to official science, any system that possesses these three distinct characteristics, will be under Natural Selection action forces. Then, how could the formation of a sequence of amino acids comply with the principle of efficient metabolism; there is not metabolism in the forming process of a protein. It is hard to explain; it seems that official science entered an open contradiction when stating such three properties to be applied in the forming process of any protein.

Other authors stated that selection is strictly based upon self-replication, so it means that when the placing of amino acids is completed, so the protein, then Natural Selection will decide if that particular protein will stay or will be negatively selected. As has been noted Natural Selection is acting twice, first during the placing process of each single amino acid, and then on the protein, when it is completed. Then again, if the sequence of amino acids were selected for the protein to be done, it is understood that the protein has been approved as a selected final product, but even so, now it will be under the action of Natural Selection to determine its self-replication ability.

Think about this, two-steps in selection just to determine if a protein will remain in the medium or not. Again, how long will it take for these two steps in selecting that huge amount of proteins which a simple unicellular organism is constructed with? I do not dare to answer this question, and in my honest opinion, any calculations made on this with the purpose of having an idea of how such processes took place in the untouchable past, is just an absurd rehearsal. Those specialists, who are trying to immerse themselves in such calculations, are missing the core of the problem: a complete ignorance of how the environment was in that era, which is in fact lost in the darkness of time. All they do is pre-suppose how those environments were, mainly based on certain, (and usually incomplete) information furnished by rocks and layers.

In any event, if the molecular selection was so strict, why so many pathological conditions exist in *Homo sapiens* (nowadays humans) with no selection at all for the healthiest? Now, on top of that, some analysts try to explain that during the last 30 or 40 thousand years modern man, again *Homo sapiens,* was under the pressure of recurrent mutations that led to the present inheritance status of so many **pathological conditions**; many of them are recognized as having a strong genetic component. In fact, some 800 genetic-based diseases are known today. If ancient humans, thirty or forty thousand years ago, did not have medical care, and of course they did not, how did they survive long enough for us to be here as their direct descendants? It was supposed that the powerful Natural Selection would send them to the brink of extinction, and that did not happen, why? Moreover, they lived in constant danger due to the fact of hunting, hiding from predators, looking for shelter under inclement weather conditions, aggressive behavior in their groups, fighting for existence, wound infections, just to mention few of the possible harsh life circumstances they had to face, in every single day of their extremely hard life.

Facing all these, simple questions may arise: Was there any Counter Natural Selection force? Do modern scientists investigate such a possibility, just to explain why so many life-threatening diseases appeared on the stage for present-day humans? Beyond that, today known mental disturbances, in many cases, are explained on a genetic basis, and they really are. Molecular biology and Genetics have contributed with important insights into behaviors connected to pathological disorders.

Chances are that some of these psychiatric conditions have been inherited from our ancestral apes or archaic hominids, and at the light of human behavioral codes they are tagged as mental disorders, being nothing but ancestral genes passed through generations in the behavioral evolutionary pattern of our incredible hominid history; then again, natural selection failed to get rid of them in the long course of thousands of years, or are they new acquisitions? We are not a brand new species; we are the net result of a long journey of evolutionary inheritance and change. If those genes are the explanation for the so-called mental disorders, and then this assumption is true, it is by no means possible to deal with such behaviors daily in the privacy of a regular modern family, since they pertained to ancestral fore-runners of modern humans, whom cannot fit into our social-accepted-behavioral-patterns. Those genes belonged to another era lost in time, and we are the children with a portion of that gene pool, no matter how good or bad it might be. Besides, we do not have the slightest idea regarding our forerunners daily behavior, their social patterns, and their intra-group relation-ships; probably we will never do.

What was a customary behavioral pattern for that/those species, it is unacceptable for us today. Is any psychiatrist in good shape to determine, or at least propose, the possible origin of the most common today mental disorders? I am afraid not. The answer is lost in time, not even the most outstanding paleoan-thropologist may approach such an answer.

Medications, chemicals, are administered to patients with such disorders to counter-act the expression of said genes, but if those genes really represent an ancestral inheritance, noth-ing could be done to eliminate the so-called disorder, just keep the expression of these genes in a very low level in virtue of the administered pharmaceuticals. Few, if any, researches have been accomplished on this topic, and these entire disturbances lie on psychiatric science, not focused as a remnant of our ape/archaic hominid ancestors. Just remember that in medieval times, *gale-nos* strongly believed that insane people were nothing but de-mon-possessed souls.

All human concepts lie on, and belong to, the epoch and culture where they arise, and they are tight to such social con-ditions and belief systems. In the history of humankind and in that of science, many things are taken for granted, and so Natu-ral Selection in our time is. Human behavior, undoubtedly, is in part due to environmental factors, but a non-meager portion of it is fully explained in terms of genetic background. Psychiatrists and psychologists perfectly know this. Some researchers try to discredit Evolutionary Psychology, using epithets just like pseudo-science; this is not a matter to ill tag a new attempt to shed some light where classical anthropological studies failed in many ways: measuring bones, digging, and trying to interpret the morphology of fossils, and that is it. It is worth to mention that evolutionary psychologists have presented more than a few interesting ideas on the possible environmental pressures that made possible language to evolve, which have been a prospective contribution for a better understanding of our evolutionary history.

We are now in the 21st century, we ought to be opened to new trails in science, and history will determine if those trails are leading right or not. Genetics, at the turn of the last century was not a science either and today is one of the leading disciplines in Biology. Many studies have been carried on primates to better understand our own behavioral patterns, but the problem is not just a mere comparative behavioral-pattern chart or conclusions. The problem may lay on a genetic approach, which is the true aspect of this issue, not a scrupulous registration of induced reactions or designed tests to compare "intelligence."

I do not want to extend on this aspect beyond molecular genetics on this chapter. There is an oncoming one, devoted to anthropological evolution where important aspects of human behavior will be approached. At this point, some illustrative examples could be provided. An enzyme, monoamine oxidase, has been proved as related to antisocial behavior and particularly, a low activity of this enzyme can lead to aggression and impulsivity; of course, there is a structural gene for monoamine oxidase. Besides, low levels of serotonin (another enzyme) account for impulsive behavior and emotional aggression. It is very possible that aggressive behavior in archaic hominids was positively selected, since this kind of behavior could assess better survival possibilities in an environment where essential resources were scarce. The more aggressive the better fit for such environmental conditions and survival.

Particularly, when mating was the issue, those aggressive individuals had an edge over the rest of the members in the group, and their offspring carrying the same behavioral patterns would be favored by natural selection. It is completely an act of ingenuity to believe that our ancestors living in hordes, under extremely harsh conditions, with famine as a daily visitor, could be pacific, easygoing people, leaving the meager piece of food for the others in the horde as a supreme act of altruism; the most probable context was an aggressive behavior to secure a portion of the scarce food. How long ago was this behavioral pattern onset? We have to agree it is very difficult to determine, unless archeologists could find an ideal site with well-conserved crashed skulls, amidst the remnants of any animal that could served as a meal, and this whole scenario could be interpreted as a crucial battle for feeding, with victims around. We cannot be surprised realizing that our archaic ancestors were not pacific individuals. Actually, the inclemency of the environment could shape the behavior of our ancestors; the weak, the shy, the inoffensive ones would succumb in the horde. Only those genes that guarantee survival would come the long way through thousands of years, and now, chances are that they are present in modern humans, becoming a real obstacle in our current social environment. This does not mean that human behavior is **exclusively** determined by a genetic endowment, it is completely dim to focus on this matter in such a simple way, and it becomes a ridiculous reductionism. Undoubtedly, environment played an important role as well. Multiple genes, like in many other traits, determine behaviors. The most probable inherited are those genes that in the long distant past determined survival possibilities to our ancestors; **evolutionary success**, in simple words, is nothing but survival. Now, it is very simple to understand why in our "civilization" strict law systems are legislated just because of inherited genes; it is this: at present our social, moral, and emotional evolution will be a painful road as long as we have such detrimental genes flowing in today human biology, inherited from our savage forerunners. A daily life experiment is very simple: while you are driving in a city or on a highway, just look at your companion drivers and you will realize what aggressive patterns could be observed in many of them. Of course, primitive hominids did not drive; we could not be here, otherwise. Many of them are threatening you with incredible maneuvers on the road, risking everything, including their own lives, just to let you know that they are the leaders on the road. This should be interpreted back in time, as our own ancestors displayed somehow similar behaviors to let other competitors know that they were the masters, the leaders of the horde. We do not have to hunt for existence, but we have to drive.

When one read certain articles written by well-known specialists, the message of those papers seem to proclaim that natural selection is just like a partial or limited truth, simply because for those authorities natural selection will act on any phenotypical trait but behavior, which supposedly is part of the whole phenotype. According to this, behavior would be the great exception in nature. It is unpleasant to read such papers with those incredible pitfalls, from such high analysts. Moreover, those papers (the authors) also ignore, in a rampant manner, the well-known gene interactions, which could be present in all these complex gene products. There are strong evidences that genes are malleable to some extend, turning *on* and *off* according to social information, and consequently influencing behavior. The existence of a dynamic relationship between genes and behavioral patterns is a fact.

Here we have some other contradictory concepts. It is said that: "our genes have more than enough to do for making us extraordinarily smart, giving us the physical talents...". For heaven sake, talents are not physical, but intellectual, that is why we speak and make technologies. Another expression that leads to reconsideration is when some specialists referred to the case of children born in a certain culture and then transferred as infants into another one, adopting the language and customs of the new country. Yes, of course they do, they are human beings and we are extremely adaptable as our archaic ancestors were, that is why we are here. But this example has nothing to do with the fact that behavior is culturally conditioned. What this example proves is that humans possess *adaptation skills* to new environments, and then again, that is why we are here.

The last drop in the full glass is the sentence: "we must strive to evolve a new ethic..." One thing is to learn ethics and good manners, and be polite, and another very different is the genetic conditioned behavior. Now, on top of that, some specialists point to new approaches naming them as "reductionism," when in fact considering such a complex matter as human behavior is, solely determined by the environment excluding natural selection and behavioral genetics from the evolutionary stage, is not a farsighted approach to the topic. We cannot forget that one of the most outstanding evolutionists of the twentieth century, Ernst Mayr, explicitly declared in his *magnum opus* **Animal Species and Evolution**, that the phenotypic expression of behavior is largely determined genetically. No one has proved the opposite in a convincing manner.

The classical anthropologists criticized evolutionary psychology and behavioral genetics just because these disciplines consider the ancestral environment the place where adaptive specific behavior evolved. In my opinion, evolutionary psychology and behavioral genetics are right. As an example, just take a modern feline, the tiger. This animal behavior is not modern; it evolved probably from hundreds of thousands of years ago. Nobody may think that all hunting abilities of this big cat are justified by the modern jungle environment. What this and other animals alike are today is nothing but the result of selected genes to guarantee survival through millenniums; natural selection is a fallacy, otherwise. What is happening in the scientific world is exactly what goes on in belief systems today: every single group thinks their members possess the truth, and the rest are all condemned. This extreme position prevents to grasp the knowledge in a swift and congruent manner. Astounding papers like Belyaev's studies and later Trut's on the silver fox (*Vulpes vulpes*) have been sometimes overlooked, albeit their research work proved the genetic basis of the tame and aggressive behavior in the aforementioned species. Belyaev and coworkers started selecting silver fox that showed themselves tolerant to human presence. After 10 generations of a careful selection for tameness they obtained animals that behaved like puppy dogs, wagging their tails, whimpering, and fond of human contact. Furthermore, as a correlated response to this selection, offspring of the selected tamed fox exhibited curled tails, spotted coats, and floppy ears. During 1970's further research was accomplished using cross fostering of newborn puppies and embryo transplant, all of which demonstrated the genetic basis of tame and aggressive behavior as well. When Dmitry Belyaev and his older brother Nikolay started these experiments on foxes, it was the time Stalin's despotic system, and also the time when Trofim Lysenko declared with absolute insolence that Genetics was a science of bourgeois origin. Nikolay was sent to an extreme labor camp, very similar to those practiced by Nazis, where he finally died. Dmitry was expelled from his job and condemned to the oblivion. When the Stalinism era was over, researches started again. These results were published in 1980 and 2001; the research on this topic continues, but biased researchers, perhaps not well informed as professionals, only quote papers and books from one country or two. Belyaev died of cancer in 1985, he was the pioneer in establishing the genetic origins of domestication, also demonstrating correlated characters (curled tails, spotted coats, and floppy ears in foxes) when selection for tameness was applied.

The aforementioned studies and results are unquestionably decisive on the arguments regarding the genetic basis of behavior. The striking results in just 10 generations of selection, not only on a behavioral basis but on morphological features too, tell us that a behavioral selection pressure will come up with other traits being affected; the silver fox is not a ***mammal exception*** for these genetic characteristics. What these results show essentially is a correlated response when selecting for behavioral traits. If all these happened with simple artificial selection, what would be expected from natural selection acting upon *Homo species* for hundreds of thousand years? Likewise, besides a strong demonstration of behavior-genetic relationship in *Vulpes vulpes* as a model of higher vertebrates, an unquestionable fact that sprang out is the hidden genetic variability present but not seen till selection acted upon, in this particular case artificial selection.

Thoughtful thinking indicates that the same pattern may prove valid for other animal species, *Homo sapiens* not excluded during his long evolutionary journey. Of course, it is impossible to know exactly what behavior patterns were customary in our hominid ancestors on a daily life basis; only guessing is possible, but because genes in the fruit fly are the same as in the mouse, as was stated somewhere before in this book, chances are that when natural selection acted upon archaic hominids, and why not, upon forms not considered into the *Homo* genus, certain morphological traits surfaced as a correlated response as has been proved in the silver fox studies. Any simple detail, any fact though plain in appearance might not be overlooked if we are looking for the essential truth.

A simple and common experience shared by those that had a chimp as a pet for a certain period, is that you can keep it as a "little child" or a "baby" usually for the first six or seven years of age, but not beyond; aggressive behavior flourish rapidly with sex maturity. We only differ 2% or less from chimps in DNA sequence. These baby chimps have been educated and raised in a very cozy and caring environment, but their adult behavior is genetically based; they will response accordingly despite the environment that harbored them. This is also known by zoo caretakers; chimps behavior change as they grow older. Genes are responsible for that sudden behavioral pattern, in such a way it is a drastic change that at eight years old the chimp will confront and may attack the person that took care of him/her during the first years of infancy. If we are talking about humans, we are talking about primates, and chimp is the closest to us. Chimp's genetic information is not capable to understand that the human caretaker deserves a better consideration. The savage survival genes are unrepressed, and the selected behavioral adult pattern is manifested in full strength. This has been that way since the emergence of the species, and this pattern (a molecular basis as it is genetic conditioned) is responsible for the survival benefit of chimps, to some extent.

Some anthropologists claim that our ancestors lived in a wide variety of environments, hence the impact of such environmental diversity differed on a geographical basis, and so the surroundings. With these assertions they try to diminish the importance of natural selection acting on behaviors according to the environment, during the process of evolutionary adaptation. In fact, no matter what different those environments could be; the need for survival under harsh conditions would be approximately the same. Struggling for existence, during the whole course of human evolutionary history has been the main dish of every day life.

Based on this, natural selection ought to be considered of a great importance and indeed a direct force, shaping the phenotypes upon which it has acted during the whole evolutionary and difficult human trajectory. The same authorities insist we have a genetic shortage (in terms of genes) that does not account for human behavior variations. Natural selection ignored this limitation keeping the survival behavior as a selected priority, since we do not have enough genes for every single behavioral manifestation. Probably, among survival behaviors, *aggressiveness* and *leadership* were selectively favored in our hominid ancestors.

It has been stated by two specialists that humans have only three times as many genes as fruit flies, which is true, and many of those genes seem to be duplicates of those in the flies. They also continue and state that we perform many activities that flies cannot, such as get married, write books, compose music, and many others, which is also true; according to these analysts human activities are not creditable to genes, due to the shortage mentioned before. Moreover, these specialists made clear statements regarding the contribution of genes in increasing brain size, which stands for our vast cultural frame. This has become the exclusive product of the interaction between our neurons (brain cells) and the environment. Now, I may say: what about if we compare humans to chimps from a genetic point of view? We only find a 1.3% of genetic difference, or so. This small figure accounts for the tremendous difference between an ape and a human being.

Chimps do not get married, or write books, or compose music, (I would like to find one for me to be most famous than Tarzan) though they are so close to us. Fruit flies are insignificant in such a comparison. According to that narrow line of reasoning chimps would not be so different from us, either structurally or biologically. They are our brothers not our cousins, we have a common ancestor, or not?

These specialists that confer a great importance to culture and disdain genes to some extent, insist that to really understand human behaviors we must focus on culture, its evolution and the possible interaction with biology. I have to say that culture is not an entity as to evolve by itself; it is the creation of humans, as painting and literature are, as intrinsic parts of the cultural scenario itself. So it seems like a closed circle of terms: humans create culture and culture creates human behaviors. Evolution, no matter what level is, is not a circling process. It is an ascending progression, irreversible, and qualitatively and quantitatively fashioned. Evolution may stop, but never circumnavigate or gets back, otherwise *evolution* would not be the correct term. No evidences exist of such arrangements. Cultural evolution does not happen *per se*, it is the product of the evolutionary process of human intellect capable to modify, change, arrange, impulse, or revert out a social and cultural process. Is it not enough to observe the misleading social philosophy of communists? They impose social rules to revert the advancement of human societies by ignoring, leading beyond verge, and trying to squeeze the essence of human beings. These authors are very confused about cause and effect. Culture might be a booster for next generations to come, but not the cause; the true cause is our intellectual development, which must be a product of evolution. Besides, they proclaim the necessity of identifying the basic mechanism for cultural evolution, when in fact the true mechanism is our own intellect, and culture develops according to the progress mankind is capable to accomplish through ages of evolution (genes implied), social and biological as well, in a harmonic fashion.

Others, deepening much more in the complex garbling, stated that behavioral geneticists and evolutionary psychologists wish to speak of evolution of certain characteristics, physical or behavioral, and doing so are not only ignoring how complex the evolutionary process is, but also confusing the concept of natural selection. Of course, the individual phenotype is the target of natural selection, which is out of question, but it is only so if at least one characteristic shows up for natural selection to act upon it. Natural selection is not a particular entity; it is a complex of multiple factors acting on a specific phenotype, through climate, available resources, exposure to allergens, resistance or weakness to certain diseases, etc. If the phenotype is incapable to survive or strive in that environment just because one single aspect is deficient (genetically deficient), natural selection will determine if the individual as a whole will stay or not. An ample discussion of all these topics in full details will be found at the end of this chapter in Recommended Bibliography.

Let us go back to molecular levels. Mutations have occurred during a long period, and a huge list of genetic diseases may be found in the internet; you will be surprised if you decide to search for such a topic. Among those pathological conditions alcoholism is being studied carefully, since it is almost certain that any individual from a family history of alcoholism is in a high risk of alcohol addiction disorder.

Mutations indeed occur, for the replication of DNA is not an accurate copying process. Instead, because of thermal motion, mutations take place to produce a mutant individual; no two human beings are identical. Every single person is unique from a genetic point of view.

There is a principle that the official science gives support to, stating that the longer a DNA sequence is, the more accurate its replication ought to be. In shorter sequences errors will accumulate in generations to come, and the original information is practically lost. If we stick to this principle as the official science does, eventually we have to admit that our DNA is too short to stay healthy, and here lies the reason that explains why so many errors (mutations) we bear that compromise our healthy condition, and our life span.

We are genetically imperfect, and this ostensible DNA-imperfection (molecular shortness) is responsible for our poor life span that only few humans feebly accomplished over the average, living beyond 110 years. So short is our life span that most human beings die long before they can accomplish their goals and projects in their mere lifetime. Why Natural Selection failed in such a crucial aspect? Did Natural Selection intervene in determining our DNA characteristics at all? In short, if Natural Selection always favors the fittest, then there has never been the "fittest DNA" in our human evolutionary history. Sharks are very lucky regarding Natural Selection, since they are considered, by almost all zoologists, as the most successful creatures on our planet at present. Did Natural Selection do a better job on sharks? Why? Beyond that, we live around 70 to 80 years on average thanks to medical advances, our life span would be shorter, otherwise; so short that comparatively to our *average body weight,* we live much less than a butterfly. I have an example readily at hand, the Monarch Butterfly (*Danaus plexippus*) as a winged adult lives around 28 days, and its average weight is 0.5 grams. If we consider 160 pounds the body weight for an average adult human, then such a human being must live around 11,000 years. Do the math. Surprising, isn't it?

Body weight is only one possible variable that we can take as a comparable feature to demonstrate our poor living status on earth; butterflies are not the exceptions. Just think how fragile a butterfly is. It is an axiom that evolution is nothing but the optimization of functional efficiency. Are butterflies more efficient than we are? Is it the reason why so a wonderful life-span evolution granted to them, and deprived us of such a gift? It seems like Natural Selection is just a tool, but not the brain. If we focus our attention on the other side, it is, if the butterfly would live comparatively what we live, then we could never see them at all. We may believe they were meant for our enjoyment, probably that is why their dimensional life-span units differ from ours; it is just like "evolution" forecasted the establishment of *Homo sapiens* on earth, and then butterflies came to be exactly in their own dimensional unit of life-span for us to see them and enjoy them, since we evolved a sensible mind to admire beauty. Then, Natural Selection made an extraordinary job, regarding humans and butterflies. But this is not the only surprising aspect of these two species. The genetic endowment of butterflies is much simpler than ours, of course it is. That simple DNA is capable of furnishing so a marvelous characteristic as life span is, in such a length, that we can enjoy their exquisiteness, which is also a prodigious of "Natural Selection," or so believe.

Now, it is well established by science that if nucleotides are present in an energy-rich condition, they will set up chemical bonds to form macromolecular chains of DNA, not being necessary the presence of enzymes to accomplish the process. The union of a triphosphate molecule (high-energy molecule) to the nucleotides (Fig. 3) accomplishes this energy-rich condition. At a certain moment, a pyrophosphate group (comprising only two phosphate molecules) is released and the single phosphate molecule that remains from the original Triphosphate is bonded to the sugar molecule. Fig. 3 is a schematic, simplified process that represents all of the above.

PURINE

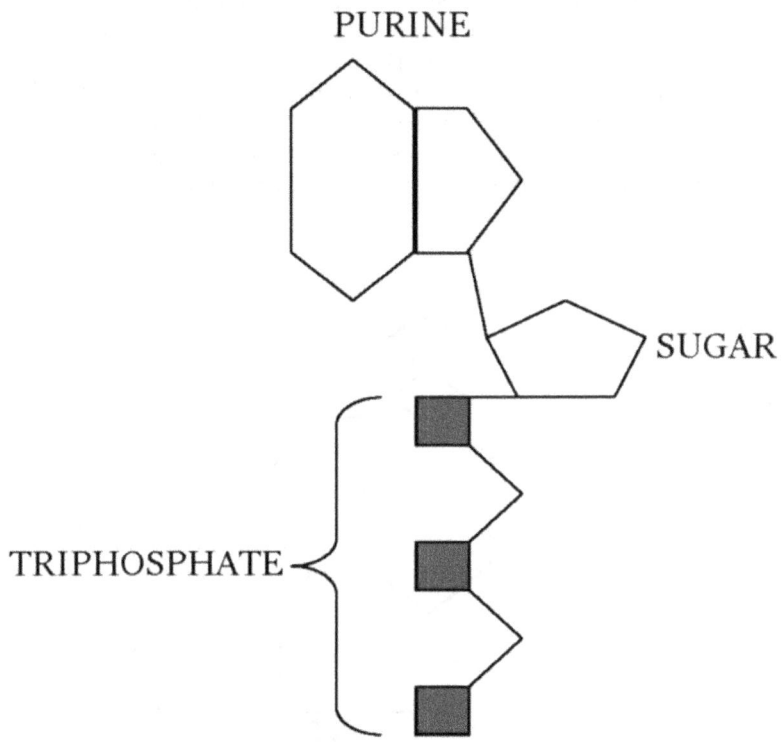

SUGAR

TRIPHOSPHATE

Fig. 3 A nucleotide linked to a high-energy chemical compound, Triphosphate.

What would happen if instead of releasing a Pyrophosphate, it would only be released a single monophosphate group (Fig. 4). It is worth to say (for those that look for more details) that a hydrogen atom (H⁻) must be released (Fig. 4), to provide and extra bond to the phosphate directly linked to the sugar molecule. If all of the above would happen, then the possibility is not just the formation of a single-strand chain at a time, but two instead. This two-strand chain eventually will bond with another two-strand chain, fashioned in the same way; as a consequence, chances are that both two-stranded chains will bond together, and the final result shall be a four-stranded DNA. In this way, the underprivileged two-strand DNA that we sport today, would simply become one of the steps in the *formation-creation* of a much better and well-finished four-stranded DNA, plenty enough for a gene-pool to overcome all present calamities of our poor DNA-dependent existence. Of course, this is not as simple as saying. For the process sketched above, a specific pool of efficient enzymes is obviously required.

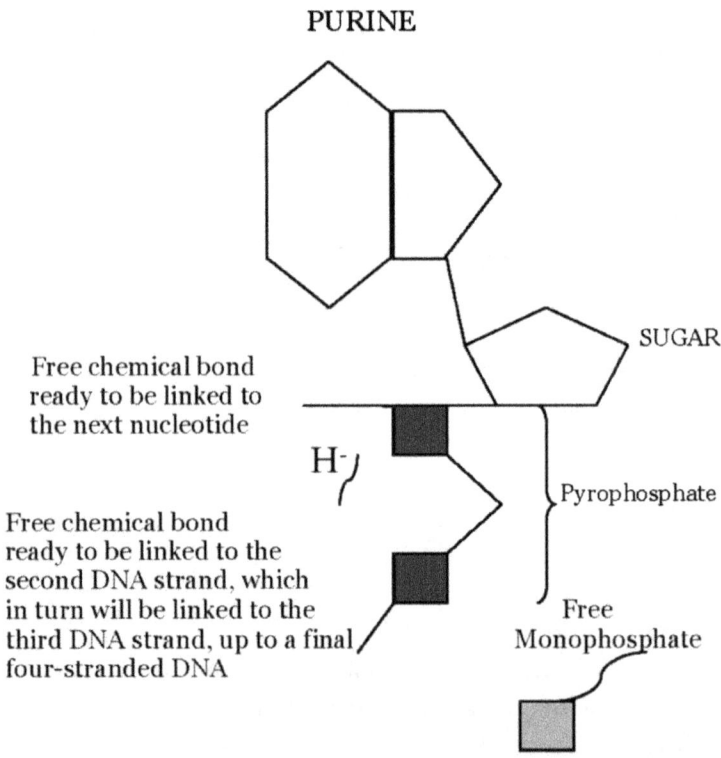

PURINE

Free chemical bond ready to be linked to the next nucleotide

SUGAR

H·)

Pyrophosphate

Free chemical bond ready to be linked to the second DNA strand, which in turn will be linked to the third DNA strand, up to a final four-stranded DNA

Free Monophosphate

Fig. 4 A monophosphate molecule is released from the original Triphosphate. The remainder Pyrophosphate group stays attached to the nucleotide to provide both the possible bond to the next nucleotide in the same chain, and the link to an extra DNA-strand.

The diagram is just an attempt or informal proposal; it is by no means a serious biochemical solution to our DNA problems. Furthermore, other mutations ought to be present to change the basic structure to some extent, mainly those connected with chemical bonds. The point is that natural selection did not, or could not, lead the formation of a rich four-stranded DNA. Then, a critical question arises: Did natural selection fail to accomplish a better DNA? Would those specific-efficient enzymes never showed up in the chemical culture of those forgotten times, or the basic structure was a hoax from the very beginning? Was it never possible for natural selection to act accordingly? Inevitably, we are sporting a very economic DNA, so cheaply fashioned that modern humans suffer countless genetically conditioned diseases, without the possible healing genes that ought to be placed as part of a well shaped and balanced four-stranded DNA, as it could be if we invoke natural selection. It is so cruel, for humans, to sport such a cheap and defective DNA in such a way that genetically transmitted diseases, from parents to offspring, would have long-term consequences in future generations to come. In this way, what a certain generation, in any family, receives as inherited genes responsible for a definite health condition will be transmitted to subsequent generations, spreading the defective DNA portion to descendants that not even knew the ancestral founders of such calamities. Then again, all of the above chemical mechanism is just a theoretical proposition, and do not intent to be an all-satisfactory solution to the synthesis of a four-stranded DNA. Next paragraph will account for real findings in the sense of multiple-stranded DNA.

As late as in the mid nineties, there were evidences of a four-stranded DNA obtained by the use of a molecular technique known as glass micro-arrays. In 2001, it was clearly demonstrated the presence of multi-stranded DNA using this technique. These multi-stranded DNA had no relation to the classic Watson and Crick base-pairing model. During 2002, several ongoing projects on multi-stranded DNA were taking place in more than one research center.

In 2006, an astounding invention was developed to immobilize the multi-stranded DNA on one micro-array. This could represent a breakthrough for the future treatments of cancer, HIV, and other diseases according to specialists. The above results up to present, may prove that the true possibility of a four-stranded DNA is not just a theoretical conjecture, but a fact beyond what was thought to be unchangeable twenty-five or thirty years ago. This is just a single and first step in the construction of a four-stranded DNA. Hopefully, more, probably much more is expected to come. What follows, is the chronology in the development of molecular research events not mentioned before in this chapter:

1958-Discovery of DNA-polymerase, the enzyme that copies DNA; 1960-Discovery of the enzyme that transcribe DNA to RNA. It is, the genetic message contained in DNA is re-written into RNA; 1960-The theoretical concept of "messenger" RNA (mRNA), and the discovery of gene regulation; 1962-Conclusive existence of mRNA. Genetic code solved; 1965-Discovery of Plasmids; 1970-Discovery of the restriction enzymes, which cut DNA in precise and exact places, together with the cataloguing of restriction sites.

At present, probably more than one scientist is really concerned about our two-strand DNA and its genetic limitations. Perhaps future science will improve what natural selection could not complete along the way in millions of years.

From Chemistry to Biology

It has been stated that selection pressure is entirely responsible for the appearance of photosynthesizing bacteria, which efficiently used the sun light as energy to create all the necessary nutrients to live and reproduce themselves. However, selection pressure is not a magic agent to make appear the photosynthesizing bacteria, no matter how long a time elapsed; bacteria are a complete unicellular organism, not a chemical compound.

The official conclusion goes that, from this kind of bacteria the cyanobacteria emerged, which also was equipped with photosynthesis possibilities. Then again, how was the transition from photosynthesizing bacteria to cyanobacteria? According to the scientific conclusions cyanobacterias, over a huge period, were able to release a tremendous amount of oxygen to start in some way, the oxidation process in our planet. Then, in a gradual manner oxygen gas spread in the atmosphere. The question is: what was the amount of cyanobacteria on our planet to replenish the atmosphere with a layer of oxygen gas? Was the planet all covered with cyanobacterias to justify the massive production of oxygen? The longer the period to accomplish that job the greater the probability of changes in the environment. Therefore, chances are that cyanobacterias could not succeed facing environmental changes; besides, we do not know the magnitude of such changes, then we are unable to express a radical conclusion about cyanobacterias replenishing the atmosphere with oxygen gas. In addition, it is a well-known principle in Ecology that the sustaining capacity of any environment would limit the expansion and success of the species connected to it. What sort of environment was that with an unlimited sustaining capacity for the cyanobacterias?

At this time, oxygen gas in the atmosphere led to the onset of the ozone layer, which filters the ultraviolet rays, which are considered of a high penetrating capacity, in such a way that all forms of life now present in our planet could not exist without the ozone-filtering layer. It sounds like a paradox to consider that, even the most elementary forms of life beneath a free-oxygen atmosphere (in this particular case the cyanobacteria) with ultraviolet rays striking on the earth surface, (as a threatening agent of life systems) could ever spread all over the planet and produce the beginning of the oxidation process in our atmosphere. Were cyanobacterias so formidably resistant to ultraviolet rays? It is hard to believe and if it was that way, then it is better to call them "ultra-bacteria."

Some openly minded scientists have stated that the so-called biosynthesis of the cell is somehow a questionable matter, but at the same time they grab the possibilities of Physics and Chemistry nowadays knowledge to explain such a biosynthesis process. Clearly, a full and convincing explanation regarding the transition from proteins and DNA to unicellular structured organisms is not at hand, since it is mandatory to include cell membranes, cytoplasm, and microstructures inside the cell as well as organelles, besides a whole metabolic system to talk about a living organism, in this particular case, the bacteria. All of this is not a simple process. What science has are presumptions, nothing else. But even so, some scientists keep on trying to explain the transition in terms of Natural Selection; they try to add more theoretical support, expressing that the fact of integration of all selected genes into a colossal genome, must occur at the moment the cell enzyme stock has such a big level of development that shall barely tolerate a very low error rate in the copying process. As a consequence, the transition from the molecular level to the cell organism is not only possible with a huge genome present in the process, the DNA replication must be marvelously **synchronized** with the cell division as well. No experiments, no findings, no scientific results have been achieved in this direction; only theoretical propositions is all we have at hand, by the moment time. As the reader may see, all of the above is exclusively based on the chemical possibility of DNA replication, but one thing is this extraordinary ability of DNA, and another very different thing is the conformation of a whole cellular organism with all the structures and internal processes mentioned before in this paragraph.

There is an aspect of unicellular organisms that needs to be discussed here. The simplest unicellular organism reproduce themselves by an asexual process known as vegetative cell division, in which the mother cell divides itself into a daughter cell; after such division is completed there is no difference between mother and daughter cell, both are young, no metabolic, not aging signs are present that could differentiate them; it is like a cloning process. This kind of simple organisms are in fact immortal. Death only can occur by misfortune, which is not the rule. Does it mean that low-level living organisms enjoy a better fate than highly developed ones, including humans? What is the purpose of such a dilemma? But such a phenomenon is not only an exclusive attribute of single cell organisms, it is well known how many common trees enjoy longevity, just as elms, poplars, oaks, not to mention sequoias. They all exceedingly surpass the average human life span. Are those trees been favored by Natural Selection in such a way that comparatively humans have been left behind? Aren't we the outstanding and superb result of Evolution? Did Evolution mislead the logical course and selected certain beings with characteristics that humans may long for?

Humans are so creative that the advancement of humanity all over the planet is the direct effort and practical intelligence of *Homo sapiens*; why an immobile tree lives so long and we live so short, up to the point that eminent personalities could hardly conclude their unselfish, astounding, and human-beneficial works? Examples of all these abound all along the human history: care-giving doctors died fairly young when more needed by their patients; exceptional musicians in the past were gone at the pinnacle of their creativeness; scientists' unconcluded studies remained for a long time in the oblivion, due to a sudden death. All of the above urge us to think that Natural Selection made an evolutionary dreadful mistake giving trees, uncreative and not rational creatures, the longevity that humans deserve to continue the improvement and progress of living conditions in our beautiful planet. Is there an answer to all of these questions beyond our logical and critical thinking? Is there something that, albeit applying Formal Logics we cannot understand as a whole phenomenon? Is it just because *Homo sapiens* is too young on the planet, hence we have not reached the knowledge to give an unequivocal answer to such embarrassing questions, in the same way Copernicus was unable at his time to solve what nowadays are simple technologies?

Besides, the point is not Natural Selection, or even Evolution that occurs as an undeniable product of the former, the most important point is: why *Homo sapiens* is here? Why is this not a planet of just plants and animals? Humans being on earth are more a disturbance for the ecosystems than a benefit. Animals never ever will harm the ecosystems since they evolved together in a miraculous coexistence. Humans do. Was it necessary for the planet the presence of *Homo sapiens*? What is the benefit, in terms of evolution, the planet receives from humans? We have exterminated many species already, the examples abound, and are notorious; the list continues and not even organizations can completely stop such savage destruction of both species and whole ecosystems. Of course, natural selection favors those characteristics that benefit the population or species, no matter if the species deploy the environment, or finally face its own extinction. Animals do not destroy ecosystems; there is a perfect balance between irrational beings and nature.

Back again to unicellular organism, in the whole context what is unsolved and remains as a vast gap is the complexity of systems, organelles, and processes, all working in perfect harmony in the cell. How could they develop at the precise time, or simultaneously, and then give birth to higher forms of living organisms? What a tremendous coordination for them to work in perfect cooperation, and make higher living creatures a real outcome; and all these extreme, unbelievable synchronized processes occurred simply under *trial and error* events, as Natural Selection does. Then again, to successfully accomplish those tremendous evolutionary steps, thousand of million years passed, and with them, deep environmental changes occurred. One of them, no questions about it, was the stabilizing thermal processes on our planet. What was certain in a specific geological period was completely different half a million years after; fossil records may account for that.

To believe that such a complex and difficult transition was exclusively led by Natural Selection (trial and error), regardless of environmental changes, is to deny the most essential ecological principles, already established and proficiently proved. What is more amazing is the fact that all the steps from Chemistry to Biology happened as a program does. It is amazing, in all those elements among other factors, the precise moment of assembly, the necessary enzymes and the programs they carry out, the emergence of mRNA, the environmental factors that account for a lot, and the harmonious structure of systems. How all these extremely complex steps came into existence just without a purpose? But...the outcome was a living cell organism. This is exactly the point: **Without a purpose?**

When facing all these aspects and questions, the most plausible conclusion is that Natural Selection is not a blind process, it acts so coordinated, and so precisely to generate life as it would be intelligent, a guided unknown "natural force" that endeavors the incredible thriving of life in thousands of forms. What selection really is? The answer is opened. But Selection alone can do nothing, as the tools of a carpenter can do nothing without the raw materials. What is then necessary for Natural Selection to act properly? The answer is simple: **genetic variation.** In fact, genetic variation may occur at different levels in the complex of heredity:

1) In the DNA strands. 2) In the genes, comprising different changes in the structure. 3) In the proteins, because of a non-exact transcription. 4) In the function of the proteins, with unexpected disturbances in metabolic or synthesis processes. 5) In the chromosomes, which the most common event is a recombination of chromosome material; through a *crossing-over* process (Fig. 5).

A B   crossing-over

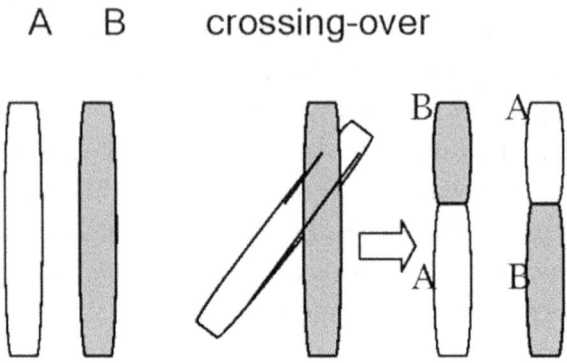

Fig. 5 Chromosomes A and B go through a crossing-over process. At the end, chromosomes have exchanged segments.

Simple variations in the sequence of the DNA molecule may turn into different gene sequences. Changes in gene structures may lead to new versions of that particular gene, which are known as *alleles*. Different alleles of the same gene will account for variations in the individual bearing them, such as body structure, height, skin color, and any other metabolic or physiological characteristic, including those of behavioral nature, too. If the transcription of the information from DNA sequence to mRNA is not accomplished accurately, then an altered protein molecule will be produced, which functionality may diverge from the original one. Changes in chromosomes are usually observed due to the process known as crossing-over, where an exchange of genetic material is accomplished as mentioned before.

Once the variation has been produced, and the consequences of such are tangible, the mechanism of Selection starts to take steps. According to Darwinian postulates, if any mutation (genetic variation at the genes level) represents a deleterious condition for the individual, Natural Selection will get rid of it during of selection. Nonetheless, today humans suffer from multiple health conditions that are determined by genes. How Natural Selection couldn't act on those diseases efficiently enough to get humans free of them, if the evolutionary human history has been about million years? Some basic reasoning on this point leads us utterly to conclude that the "perfect genome," the *non plus ultra* of evolution-creation phenomenon, is the genome with non-defective mutations. It does mean: perfect to strive in its environment since genome and environment have been and shall be a duet. But environments change: those changes are just like a "punishment" for the "original sin." Hence, the perfect genome is no longer a possibility in such changing environments; the *Garden of Eden* is prone to change, without previous notice.

At this point, it is necessary to enter in a very special aspect regarding genes. Before 2008 it was believed that 50% out of the 25,000 genes identified in the human genome, were silent or non-functional; no activity was detected in those genes. It was really odd, it would seem that such a huge amount of genes were there in human chromosomes just for the fun of it. It was totally upsetting and difficult to explain in terms of biological efficiency. In 2008, it was discovered at Colorado State University, that the "silent genes" in fact have a rhythm that is perfectly detectable. Those silent genes were isolated and it could be established that they express themselves in a low level, being coordinated with the rhythm of expression of active genes. They play a role in our bodies, and this has immediate consequences in our biological systems. It is believed that they could influence on the aging processes and chronic diseases, as well. Now, the door is opened and chances are that soon some explanations could be at hand, to better understand what in the past seemed to be unanswerable questions about human genome. Are those silent genes really connected to the aging process in humans as it is suspected? Are they the clue to answer the question about our short life span, mentioned somewhere before in this chapter? Do they need be more active to really be effective in lengthening our life span? Or will they actively act on chronic diseases to ensure a longer lifetime?

These and other questions shall arrive based on this recent discovery. Human genome stays to be somehow a modern mystery, no matter what our knowledge is at present. It seems that very delicate imbrications do exist in the human genome, making it a factual highly integrated system, far beyond the conceivable status of knowledge hold by the most eminent scientists at present days.

HOX Genes: conserved throughout evolution

Hox genes (Homeobox genes) are located in clusters, and the proteins they are encoded to manufacture are the paramount master regulators of all embryonic development in every single species studied to present, including plants. Beyond that, the expression of these genes is extended all the way through after birth. All indicate that Hox genes have been conserved throughout evolution. At the end of the 1970's, it was determined the relationship between their chromosomal array and their physical expression in the phenotype. Mammals and birds have four identified clusters of Hox genes in their genomes; plants and unicellular organisms do not exhibit Hox genes in a cluster fashion. During the formation of the embryo, Hox genes determine the specific development of biological structures, according to evidences obtained from recent studies carried by different research groups. It is evident why Hox genes linger through evolution, since they determine the embryogenesis process, in a very accurate and coordinated manner. Nevertheless, mutations occur in those clusters of Hox genes, and diseases as cancer have been correlated to mutations in certain Hox genes in humans. Once again, the imperfections of DNA transcription generate such mistakes as to create threatening-life diseases as cancer is. This is another proof that our DNA is imperfect, erratic, and liable for ruthless mistakes during the transcription process, which in turn yields undesirable results in any biological system. Most important, it seems like an inexplicable contradiction that natural selection has overlooked such an imperfection in the most important molecule of living organisms, the DNA. Despite millions of years of evolution, after the perfection of the living matter that has reached peerless domains, and the advent of beings with consciousness and rational minds, we find at the end that natural selection completely failed to discriminate improper transcriptions of single genes as well as those included in Hox clusters, just to make us miserable in our genetic endowment. We have no other choice that to declare the imperfection of natural selection; imperfection cannot create perfection, hence, our cheap DNA with all the untold consequences we are suffering in our human lives.

The Neutral Theory: natural selection is of a second importance

During 1965, Zuckerkandl and Linus Pauling discovered that amino acid substitutions occurred at a constant rate. They obtained these results when comparing the amino acid sequences of hemoglobin from different species. They also found that the rate was surprisingly high. A short time later when Motto Kimura from the National Institute of Genetics in Japan, read the published results of Zuckerkandl and Pauling, immersed himself in the elaboration of a new theory, which he announced in 1967: *the neutral theory*, also known as *the neutral mutation random drift hypothesis*. This theory states that most of the changes or substitutions at the molecular level are nothing but a ***random fixation process*** through ***selectively equivalent*** mutants, which are the product of a continuous mutation pressure. The central structure of Kimura's theory is: the amount or extent of variation in a population depends on the rate of mutation and the effective population size. The ***rate of accrual of genetic variation*** is precisely the core of the neutral theory. Besides, as different molecules have different rate of mutation, certain implications were considered:

First: different genetic loci in the same individual will build up changes at a different rate.

Second: the same gene shared by different species will show the same rate of mutation.

Third: important molecules (for the organism) will change
at a slower rate than the unimportant ones.

The importance of the neutral theory lies on its position
regarding Darwin's natural selection theory. Both are evidently
opposite approaches to the evolution phenomenon. Subsequent
studies proved that Kimura's theory is reasonably correct at the
gene level. In 1991, Kimura published further evidences that
support his theory. It is worth to say that Kimura put forward
that neutral evolution is responsible for most, but not all, gene
changes. At present, biologists accept almost unanimously, that
the variation of genes is a wide-ranging occurrence, besides being produced randomly.

The main aspects of the neutral theory have been explained
in as much details as possible in the previous paragraphs, since it
is the most plausible alternative to the Darwinian natural selection theory. In fact, it is not only a simple alternative but also a
drastic opposite issue focused on the principle of randomness
variation. But even so, the so called random drift event of gene
variations (it is, mutations), does not seem to be exactly at random, since *ad hoc* variations always appeared to bring into being
the necessary changes that lead evolution in the proper way, not
only in the emergence of new species, but in the enormous biological diversity that has populated our planet ever since.

Furthermore, the content of the **first** implication mentioned before, lead us to immediate questions: based on what
principle the different rates of variation are adjudicated to the
different genes in the same organism? What natural law or process determined the diapason of rates to act on the suitable
genes? Is it all explained ingenuously by random variation?

Now, the **second** implication states that certain genes will
fit different organism and those genes will reveal the same rate

of mutation, regardless of the organism bearing them. Walter Gehring in 1994 demonstrated that the gene that determines the eyes in the mouse when transferred to the fruit fly, this insect will develop normal compound eyes as any other fly. This experiment proved beyond any doubt, that same genes are shared by species wide apart in the evolutionary tree and according to the second implication those genes will hold their rate of variation immutably, no matter which organism bears them.

The **third** implication is strikingly controversial. What force or natural law determines that the most important genes for the organism will show a low mutation rate? How this thing can be decided at the molecular level? What is beyond such an important decision?

In fact, if we take an extremely closer look at the three implications, we will be facing somehow similar questions and dubitative situations as we found before when dealing with the postulates of natural selection. As a colophon, we may say that something is unclear and unsolved in the "evolutionary process," as long as the explanations, both natural selection and neutral theory, are based exclusively on theories that move around the concept of, *living organisms subject to the will of nature.* In this way, we are immersed in a tautological runabout that shall always lead to solve just certain aspects, being others left behind.

Mitochondrial DNA and the Mitochondrial Eve

In the study of DNA not only the cell nucleus is a target of most interest, there is an organelle in the cell of multicellular organisms by the name of mitochondrion (plural mitochondria), being its main roll in cell biology to provide energy for the cell to use; hence, these organelles are also known as the *powerhouses* of the cell. Besides, mitochondria contain DNA molecules (mtD-NA).

Today, we know that all mitochondria are *inherited from the mother's gamete or ovum*. It has been proved that most of mitochondrial DNA mutations show the way to functional issues; therefore, mitochondria is crucial for higher organisms to keep up living. It goes to such extend that the slightest difficulty with any of the abundant proteins (enzymes) used by mitochondria is devastating to both the cell and the organism itself. As the reader has been acquainted, the inheritance of mitochondria is direct from the mother. No logical explanation, or at least an accepted one, is available regardless of researches carried on. Now, in 1987 biologists from the University of California, Berkeley, and from the University of Hawaii, Honolulu published a paper from a research that included 147 individuals drawn from five geographic populations, and the results showed that all those mitochondrial DNA stem from one woman that probably lived between 140,000 and 200,000 years ago, most likely sited in Africa. This archaic woman has been known as the **Mitochondrial Eve** as she was defined as the matrilineal (maternal lineage) most recent common ancestor; for all nowadays-living humans. It does not mean that the Mitochondrial Eve was the only living woman at the time. It means that tracing back the matrilineal descendants they arrived at a woman that, among others of her own population, represents our most recent common ancestor in the maternal lineage concept.

The analysts suggested that this woman belonged to a small population, perhaps no more than 10,000 individuals, perhaps less. What does evolution intended favoring the inheritance of mitochondria only from the females in any human population? According to Darwinian Theory, a character is positively selected only if it enhances the fitness of the individuals carrying it. Then, from a Darwinian point of view the question is straightforward: How is this kind of inheritance (matrilineal) beneficial to human populations? No answer yet. Perhaps, the most plausible one is connected to the possibility of tracing back our developmental history, which is lost in the darkness of time. Then again, what is the direct benefit of such possibility in terms of Darwinism? This is another question, with no answer. Among biologists it is a common criterion that probably today mitochondria were free-living bacteria billions of years ago. These bacteria developed a symbiotic relationship with their host, becoming an integrated part in the host's life cycle ever since. According to classical studies it is accepted that primates, humans included, carry some 1,700 mitochondria in every single cell.

According to molecular biologists, mitochondrial DNA mutates ten times the rate of nuclear DNA. Nevertheless, these mutations are "neutral," so natural selection does not generate any action on them. It is comprehensible, since if mutations subjected to natural selection already occur in the nuclear DNA, then if they would occur in mtDNA with no neutral effect, then, a double mutational process could affect the genome as a whole with too much variation load. Hence, the neutrality of mutations could be another asset in using mtDNA in all these kind of studies. In fact, mtDNA is a very simple and accurate genetic marker, since it is possible to trace back a long chain of generations, based on it. Above all, mitochondria can be found in almost all living organism: plants, insects, fish, amphibians, reptiles, birds, and mammals.

It was Douglas Wallace the first scientist to prove that mtDNA was a maternal lineage inheritance. Later, other group of specialists determined that human mitochondria mutate at a rate of between 2 and 4 percent in a million years. Of course, other scientists oppose to the results obtained through mtDNA; mainly those who stick to the classical paleoanthropology discipline. One of the critics against mitochondria is the fact that a "biological clock" determined by this technique is by no means reliable, since it is based on a steady rate of change (mutations), which has not been completely proved.

One thing is clear, classical paleoanthropology bases its results on physical measurements and shapes found in the fossils, which conclusions may differ from one researcher to another.

A portion of those conclusions rests on conceptual interpretations, hence the different approaches to the same problem. Not long ago, it was accepted that Neanderthals were one of the links in human evolution. Nowadays, it is the opposite, thanks to molecular biology, not of conceptual interpretations based on physical resemblances. I do not try to say that molecular biology must replace paleoanthropology; the true scientific and logical procedure should be both disciplines working together, and let the results prove the rest; but that is not exactly what has been happening throughout the research work for many years. Every single specialist believes to have the truth contained in his/her results, not realizing the convenience of complementary results from different points of view, and then integrate them all in a consistent, coherent, and convincing conclusions.

What is inside in all this issue could be explained in a very simple manner. No matter what the technique or discipline they are applying, many gaps do exist in the process of evolutionary trees. The only handy tool to fill up those gaps is a little dosage of guessing, in one field or another. It is precisely here, where discordances surface, becoming this problem not just a matter of *scientific proof*, but also a great bout of interpretation and speculation to some extent. If it were a matter of scientific proof and nothing else, there would be no problems at all, since every single scientist would be satisfied with a true *scientific proof*, just because they are scientists. Besides, when a certain topic has been approached with different techniques, different scopes, including different research groups and opposing criteria still stand out, it appears that such an issue is not very clear yet; no opposing interpretations will float up, otherwise. Further investigations are mandatory, until a convincing consensus may flourish, unequivocally. But this is not exactly what usually happens with all these disagreements, every research group try to establish their own *status quo*; the projected image of such discordance to the non-specialist reader, which is the majority, is that no one is really grabbing a solid criterion on the issue, and worse, the truth seems too far from those analysts and their methods, reasoning, and applied technology.

One particular criterion seems to be shared by most of the specialists nowadays involved in the study of human evolution, it is this: if there are strong evidences that about 1.8 million years ago, there were four hominid species sharing the same area in Kenya, though the region was vast enough, why is *Homo sapiens* the only hominid species alive today? Some explanations have been tried to answer the critical question. Among others, cave art, carving, engraving, notation, ornamentation, and burials as intellectual advances, may explain what we are today. To answer the aforementioned question in this manner, produce the logical outcome of other questions: what really happened that all these intellectual manifestations took place? Did the other hominid species reach the same advances? Were our ancestors so genetically different from the rest of the other three hominid species that the latter disappeared and we are here? How could this be possible in a vast territory that comprised part of Asia, Europe, and a considerable territory of Africa after some groups left Africa?

The populations of all four species were scattered in that vast territory, so no scarcity of resources could be claimed to explain the three species uncanny extinctions. To sum up, it appears that at present there is not a convincing explanation for the unprecedented survival of our species, *Homo sapiens.* What does really lie behind our incomplete present knowledge regarding those obscure and blurred events of yore? What exactly classical anthropology and evolutionary genetics cannot decipher from the past? *Mehr licht.*

Recommended Bibliography

Belyaev, D.K. 1969. Domestication of Animals. Science. 5(1): 47—52

Belyaev, D.K. 1979. Destabilizing Selection as a Factor in Domestication. J. of Heredity. 70: 301—308.

Cann, R. L., Stoneking, M., and Wilson A. C. 1987. Mitochondrial DNA and Human Evolution. Nature 325: 31—36.

Carroll, Ch. 2010. Hubble Renewed. Nat. Geog. Feb. page 122.

Clayston, J., Dennis, C. 2003. 50 Years of DNA. Palgrave McMillan. N. Y. 144p.

Cavalieri, L. F., Finston, R. and Rosenberg, B. H. 1961. Multi-Stranded Deoxyribonucleic Acid as Determined by X-Irradiation. Nature 189: 833 -834.

Ehrlich, P. R. 2002. Human Natures: Genes, Cultures, and the Human Prospect. USA, Penguin (non-classic) 544p.

Eigen, M.1992. Steps Towards Life: a perspective on evolution. Oxford University Press. N. Y. 173 p.

Forum in Anthropology in Public. Genes and Cultures.2003. Current Antrhopology 44: 87-107

Kukekova, A. V., Trut, L. N., Chase, K., Shepeleva, D. V., Vladimirova, A. V., Kharlamova, A. V., Oskina, I. N., Stepika, A., Klebanov, S., Erb, H. N., Akland, G. N. 2008. Measurement of Segregating Behavior in Experimental Silver Fox Pedigrees. Behav. Genetics 38:185—194.

Lappin, T.R.J., Grier, D.G., Thompson, A., and Halliday, H. L. 2006. HOX GENES: Seductive Science, Mysterious Mechanisms. Ulster Med J. 75(1): 23-31

Lewin, R. 1999. Patterns in Evolution. The New Molecular View. Scientific American Library. N. Y. 246 p.

Mc Inerney, J. D. 1999. Indications that Behavior has a Biological Basis. Foundation for Genetic Judicature. Genes and Justice. Vol. 83 (3).

——————————————. 1999. Genes and Behavior. A Complex Relationship. Judicature. Genes and Justice. Vol 83 (3).

Meacher, M. 2010. Destination of the Species-The Riddle of Human Existence. O Books. John Hunt Publishing Ltd. 244 pp.

Ratliff, E. 2011. Taming the Wild. National Geographic. 219 (3): 34-59.

Robinson, G. E. 2008. Genes and Social Behavior. Science 322: 896-900.

Shi, S. J., Scheffer, A., Bjeldanes, E., Reynolds, M. A., and Arnold, L.J. 2001. DNA Exhibits multi-stranded binding recognition on glass microarrays. Nucleic Acid Research. 29(20): 4251—4256.

Tattersall, I. 2001. Evolution, Genes, and Behavior. Zygon 36: 657—666.

Trut, L. N. 1980. The Genetics and Phenogenetics of Domestic Behavior. Proceedings of the XIV International Congress of Genetics. V2, book 2 (123—136). Edit. MIR.

Trut, L. N. 2001. Experimental Studies of Early Canid Domestication. The Genetics of the Dog. CABI. P 15—43.

Trut, L. N., Kharlamova, A. V., Kukekova, A. V., Acland, G. M., Carrier, D. R., Chase, K., and Lark, K. G. 2006. Morphology and Behavior: are they coupled at the genome level? The Dog and its Genome. Cold Spring Harbor Lab. Press. Woodbury, N.Y. p 81—83.

Chapter 2
Genetic Variation. Is it the exclusive raw material for evolution?

The most impressive attribute of living things, species, is **variation**. Life, as materially understood, is inconceivable without variation. Humans, perhaps without knowing it, emulate nature as long as they build things with a high degree of variability. Just take a look at any human design: furniture, cars, computer programs, architecture, and art in all its manifestations, just to mention a few. They all proudly show a tremendous amount of variability in many ways. It is just basic logic to interpret humans' first creative learning, as constructing things, to imitate nature variations.

As it was clearly stated in the previous chapter, genetic variation starts at the molecular level, for DNA is not copied without faults; these are responsible for the variations inherited according to simple Mendelian inheritance and multi-allelic traits. Of course, when we accept all the perfection that life implies, it is somehow uncanny that such an exquisite and well-made molecular structure as DNA is, proves such a disadvantage as defective replication. But at the same time it is admirable that a conceptual replication mistake becomes the *sine qua non* of basic biological variation. Was it just a mere coincidence that the simple and somehow inexplicable replication error became the corner stone of evolution? In fact, there are strong evidences that evolutionary divergence rests on strict differences in the order in which chance variations appear, not on different directions of selection. What comes out is that the importance of chance variation is not less than natural selection.

The **central principle of life is variation**; the living world would be a repetitive exact-copy-process of the same models through the endless discourse of earthly times, otherwise. Then again, is the so-called defective replication of DNA a true and testable flaw? Or is it an amazing "directed" phenomenon to make life possible on earth? Simple in its grandeur, is the treacherous DNA copy; it is the time when a mutation occurs. But not only mutations are responsible for genetic variation, the so-called *gene flow* is also important in providing genetic variations. In vertebrate animals, humans included, gene flow occurs when individuals from one population migrate into another population, usually nearby, and if the genes they carry are not present in the receptor population, then new genetic material will be an acquisition for the receptor population. Furthermore, as mating with newcomers would be a fact; new gene combinations will be possible. For the reader will be clear that *mutations, gene flow* and mating new comers will produce some *genetic shuffling*, which will create new combinations of gene pools. These three events are plenty to cause genetic variation. It is worth to say that probably one of the first important roles of variation was to face different conditions and sudden environmental changes. Without such genetic variability it would not be possible for species to evolve, due mainly to the changing environmental conditions of all imaginable combinations, including but not limited to radiation level, temperature instability, resource scarcity, humidity, extreme competition, lack of proper reproductive setting, lack of appropriate shelter, etc. The only strategic bio-defense against all of these disasters was genetic variation, *ab origine*. When did genetic variation appear for the first time in evolutionary history? It is not a simple question, but according to the principle that without variation evolution cannot take place, the most plausible assumption is that variation happened really early among the primitive living forms, and then continued to the present. In any population not two individuals are identical, humans included, except for monozygotic twins, who in fact show some slight differences easily detected by their parents and relatives; hence, they are not identical either.

The expression *genetic variation* is not only applied to different organisms that evidently are biologically apart one from the other, or from different populations; it is commonly and usually applied to individuals in the same population. Then again, in the structure of any population, not only human but animal populations as well, not two individuals are identical from a genetic point of view though they may seem extremely alike, as is the case in any animal population in the wild. I may call the resemblance of individuals in any wild population as an overall resemblance, simply because if we take a closer look to a sample of individuals in any animal population, we will soon detect slight differences among them, so evident are those differences that we can tell them apart, counting on well-trained eyes. Besides, genetic variation is not only limited to external features, this variation is also observed in molecular structures as proteins in general, including enzymes of course. A meticulous analysis, mainly at the molecular level, will demonstrate that an ostensible variability is present; hence intra-population variation is the rule, not the exception.

This characteristic makes possible that if environmental conditions change, in any possible way, there are individuals with specific gene endowments capable to face the environmental changes and get along with the new surrounding conditions. Probably, not all the living elements in a population may survive drastic environmental changes, but some will do, and these are the ones that will contribute with descendants to the next generation. Those drastic environmental changes acting negatively against certain genotypes (genetic types of individuals in a population) are the operating tools of natural selection. The individuals powerless to overcome the changes in their living environment will succumb eventually. Those genotypes will no longer contribute with descendants to the next generation. This is the simple explanation for **natural selection action**. From this point on, all the implications of genetic variation as a magnificent biological phenomenon, will be discussed and analyzed along the chapter.

In any population of plants, animals, and humans individual variation is the common denominator of them all. Every single individual in any population possesses a genetic endowment, which is unique. In a flock of any species of birds, as an example, all individuals may look quite similar, but a closer observation will provide slight differences that are known as *individual variation*; these variations guarantee that any mating pair will produce offspring, which will be bearers of genetic variations as well. As has been stated before in this chapter, without such variations evolution could not be possible, simply because natural selection acts upon genetic variations present in all individuals of the population; some variations will be favored others will be ruled out. Now, it will be very simple for the reader to analyze and understand the sequence of factors that will flow out in the existence of species, humans included. The basic sequence of factors is as follows:

genetic variation $\longrightarrow$ natural selection $\longrightarrow$ evolution

Each one depends on the other on a specific basis: evolution is not possible without natural selection, and natural selection is useless without the presence of genetic variation. How was exactly possible that these three gigantic phenomena became intertwined in such a logical and strictly dependant fashion? Natural selection is marvelous mechanism acting upon any important characteristic, as long as it is genetically based; as the reader has already understood, the exclusive result of these incredible interactions is the evolution itself. But now the triad-sequence evidenced in the previous paragraph is much more than any individual factor herein considered. And then again: how this extraordinary productive three-element sequence came to be? I think is too ingenuous to believe that by just mere chances things happened, in such a way that this sequence came upon the living world out of the blues, and then events started to go on. It is as ingenuous as believing that a simple skiff came to be just because a bunch of slats came together and as millions of years went by those slats got tighten together, then sunlight, humidity, and other environmental factors, as many as you want to consider, made the rest.

To the eyes and thoughts of those that only accept strict material development, ultimately everything would come to be just because certain conditions existed in the very young planet Earth, and thousands of million years were more than enough to explain anything possible to assure life, and any other mechanism connected with it. Those analysts must know that concerning material development it is necessary to add as follow: *anything that exists independently of our action and will is matter.* In this way the concept of matter as a category, goes beyond the simple and ridiculous concept of "material" as any tangible object.

As it was expressed somewhere in this book, natural selection is not an individual entity by itself, it is just the outcome from a cluster of diverse factors in nature, acting in different ways and sometimes independently, according to physical laws, geographical characteristics, weather conditions, depletions and replenishments of any conceivable resource in the existing environment, just to mention a few. The core of the problem is the existence of *genetic variation* for the living things to respond, in different ways, to the above mention cluster of diverse factors. In this way, genetic variation is the precise and specific phenomenon for these factors to act upon, which result is the survival or extinction of certain individuals in the population, or populations as a whole. Finally, those that were positively selected will continue the lineage in an ascending fashion hopefully to better forms, making possible for the species successfully continue the course of evolution, albeit many species on Earth have almost no change for millions of years, chimpanzees for example. Once again, it is amazing to realize how so many elements are linked so well with each other, regardless of their own complexities.

Not having a better and concise explanation many specialists use the resource of time: "taking into account thousands of millions of years, anything could happen." Answers like this, on the average, only demonstrate that convincing explanations are missing and only guessing is possible, since testing is out of the question. We are dealing with tangible factors but unreachable ones; this duality prevents any attempt of experimentation to prove causality in the above-discussed topic of: genetic variation-natural selection-evolution.

It is important to consider at this point an argument that has been on the theoretical evolutionary issue for many years, and continues. A significant proportion of well known biologists do not agree with the postulate that evolution is mainly or exclusively attributable to natural selection, since the change of the genetic make-up of a population may occur not by the direct action of natural selection but by chance, which is known as *genetic drift*. Besides, it is considered that this phenomenon is prone to happen only in isolated small populations, and precisely due to the small number of individuals in those extreme populations, any change in their genetic endowment may generate a *sudden* and *drastic* modification, leading to the emergence of new species in a relatively short geological time. In support of this, fossil records firmly stand. Then again, at this moment the idea of natural selection being all sufficient to explain evolution has not been shared by many biologists and paleontologists as well, since many traits have proved to be non-adaptive at all, showing completely neutral response. Besides, other modes of possible evolution mechanisms have been on the stage of controversy for many years. The resulting conclusion of all this is that perhaps the way evolution appears to the present human eyes is not well understood whatsoever, and discussions frequently arise as a product of inconclusive criteria, certain dosages of speculations, and finally a poor or incomplete understanding of the process in its essence. Nevertheless, natural selection stands as the most plausible answer to the issue. It is somehow amazing that a single man, not well trained in biological sciences as Darwin was, could shed so immense light on the process of species evolution, giving the world the most outstanding document to approach the basic understanding of life evolution on earth. In contrast, the whole team of evolutionists during the 20th century was engaged in controversies, hard discussions, and in many cases inconsistencies regarding mechanisms, modes, and interpretations on the evolutionary issue. As a colophon, later in this chapter all the theoretical frame and evidence of sudden emergence of species will be discussed.

Geographic Variation. The Space Evidence.

Geographic variation is in fact a group variation. Physical or biological barriers, or both, usually isolate a specific group (population), from other groups of the same species. The individuals living in a certain location are facing the typical environmental effects of that specific location; hence, the factors acting on that population are molding, through natural selection, the genetic pool of the whole group, then, the individuals that are exposed to the pressuring factors of that specific environment form a group. In this way, a population is molded as a whole, depending on the specific environment its individuals are living in, which may be very different to the neighboring environments, no matter how near those environments are.

One question is important to answer, for this is the central point of the geographic variation issue: Are the physical, behavioral, physiological and structural characteristics of a population sited on a specific geographic location, genetically based? The answer is yes and no. Most of the variations among populations of the same species located at different geographic scenarios are genetically based, others are not. Those that are determined on a genetic basis are in fact the elements that support the concept of the *isolated population*, which in turn could be considered as an incipient new species; this principle is mainly applied to sexually reproducing species, and certainly humans included. If any population is isolated, this only means that a first condition has been met for a possible true new species to appear; two other important requirements have to be met, as well.

First, that isolated population would be unable to produce viable offspring if by any chance individuals of neighboring populations get in contact. Second, if viable offspring could be a possibility, then a behavioral mating isolation has to be established to prevent reproduction with individuals from neighboring populations, in case they get into contact. Then, and only then a true new species is on its way. As the reader may realize, the concept of geographic variability is of the most significance regarding the question of speciation. The *second question* in order of importance is as follows: What characteristics of a species are connected to geographic variation? The most plausible answer is: those that are intimately related to the fitness of the individuals in the group. Remember that genetic traits are the ones under the action of natural selection, and natural selection will act upon those traits that are meant to be of a survival and/or of reproductive value, in one way or another. Those characteristics that are not genetically based exist as a response to the environment, and may include responses to climatological characteristics and components of food resources, as the most common ones.

It is worth to say that about eighty years ago, scholars found that sometimes some individuals of a population are migratory since others are not. At that time studies were conducted on birds, but chances are not only populations of those birds show that characteristic of semi-migratory populations. It is by all means possible that primitive hominids showed the same tendency; some stayed and others migrated, generating a disruption in the population, and then geographic variation appeared as long as they settled down in different locations, perhaps far beyond any possible secondary contact. This particular aspect will be dealt with in the chapter devoted to all the aspects related to human evolution.

In any case, the problem of partly migrating population is not solve, since it is of a behavioral concern and it has to be precisely determined if such a behavior is just a response to the pressure of the environmental elements upon the population, with some individuals having the capacity of taking action to avoid such an environmental pressure, or is a genetically based response to protect those migratory individuals of any menacing changes in their basic environment. At this point, the reader may recall the well-known cases of the suicidal whales that abandoned the group to die on the shore. It is another example of migratory behavior not well documented or understood. Of course, regarding those unfortunate whales the behavior is of extreme consequences. Besides, remember that whales are not fish, they are mammals, the same zoological class as humans.

Polymorphism

It is a genetic variation of discontinuous characters, where one or few genes are involved in the expression of different morphs; this specific variation is always referred to as an intra-population occurrence. Any discontinuous character may show a polymorphic condition, just as behavioral, physiological, structural, and biochemical (see Fig. 6).

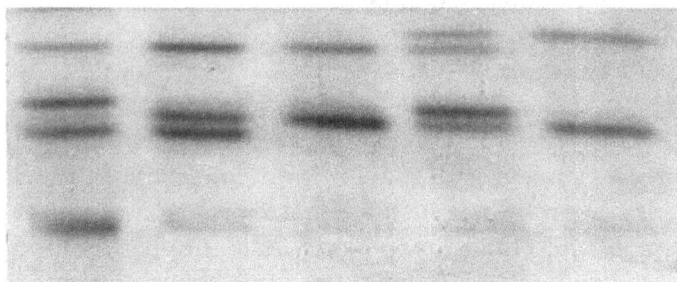

Fig. 6 Polymorphism of whey proteins detected by means of electrophoresis technique. Samples of five individuals from the same population. Each band represents a single allele. (From author's Ph. D. Thesis Work, 1981)

In general, any population is finite, but large at the same time; the biologic-population commands act in a slow fashion. A different process occurs with these commands in isolated organisms. For this reason, the supra-organism structures just like populations and species are more stable than the individuals comprising them, which perish with relative rapidity. Nonetheless, the stability of all these group structures is only possible with the outcome of new organisms, as real substitutes of the old ones by the reproduction process. Likewise, all new individuals will be bearers of a genetic endowment that guarantee survival and reproductive reactions. The genetic endowment (DNA) constitutes the genotype inherited from the progenitors. Consequently, the sexual reproduction of any organism could be affected by the interference of mutations, which in turn could affect to a lesser or greater degree the so-called bio-directive system (genotype and population gene pool), which produce changes in survival responses, and then variations in the organism-environment interactions. In addition, all these changes introduce deviations in the survival probability of the individuals. Based on the information process together with gene flow and genetic shuffling, natural selection acts in many ways to generate the evolution and emergence of species.

Environments determine the possibilities of a genotype. On the other hand, phenotypes compete among themselves for the reproductive success, which defines at the same time the adaptive value depending on the interactions with environmental factors, such as: predators, competitors, pathogen agents, and other selective pressures that vary seasonally, through the years, and geographically.

If it is considered that the genotype has a biological labor by offering an ample range of survival and reproductive responses, and that the loci are conceptualized as huge nets of interacting systems, then the indirect relations of all genes regarding their secondary effects will guarantee that all allelic substitutions must show very different effects, according to the specific genotypic combinations accomplished, as well as multiple pleiotropic (gene interactions) effects, in any possible combination. Based on all events the non-neutralist theory of the biochemical genetic polymorphism is supported, which is the manifestation of different morphs of the same protein or any other polymorphic system, as is the case of hemoglobin among others, which demonstrates to have more than one form.

If we consider that the presence of a particular gene in a population linger in spite of the opposing trends of all natural selection factors, it is not a non-sense to infer a non-neutrality condition for that gene. It is more thoughtful to consider the neutral polymorphism condition as a short-lived state regarding the genetic variation phenomena, since there is no way that natural selection may ignore the polymorphic condition in any of its manifestations, though it might be a subtle advantageous or disadvantageous condition of the morph, according to the environmental setting at the time during the settle down of the morph itself. Furthermore, the neutrality of a certain polymorphism could be partially expressed, since one or two or more alleles of the same gene could be in a neutral condition for a certain period, while other alleles of the gene are under particular selective pressures. This *status quo* may change according to the changing conditions of the environment, and then a reverse scenario may happen turning the former neutral genes into a set of selective-pressure-sensitive genes. This leads us to the concept of a dynamic polymorphic system, which of course is nothing but a cushion for the population, since it is in a good shape regarding genetic endowment as long as it has abundant polymorphic systems to face and fit to the changing environmental conditions, all along the survival process of the species-population complex.

Albeit abundant research has been accomplished on biochemical polymorphism covering many animal populations in the wild, and domestic animals as well, it is not completely clear all of the complexities and genetic potential present in biochemical polymorphic systems. Nevertheless, it has been put forward that not all the polymorphic systems studied up to present demonstrated to be neutral; on the contrary, they seem to be associated with certain characteristics of the environment, though the consistency of the association is not always sustained or proved and statistical analysis fail to be complete and sufficient. Furthermore, pleiotropy (the action of a gene affecting more than one characteristic in the phenotype) is a reasonable possibility to explain the maintenance of biochemical polymorphic systems; this aspect has not been well studied, which is mandatory before any attempt of a neutrality approach could be invoked to judge this kind of polymorphism. Besides, based on many studies comprising different animal populations, and human populations as well, the presence of multi-allelic systems in proteins seem to be the rule, which point to the fact that a different consideration other than "neutrality" might be thoughtful, since such an extended phenomenon as protein polymorphism is, could not be explained by simple approaches as neutrality, but a more convincing explanation as the dynamic polymorphism, stated before in this chapter.

The permanence of a gene in any population is subject to an unstable equilibrium, which means a constant disposition to move toward higher states of its equilibrium, as well as the essence of the balanced selection described as responsible for keeping the genetic polymorphism, already proved more than thirty years ago. As it has been discussed before, genetic variation constitutes the first step of the biologic-evolving process. As genuine biological processes mutations and gene flow together with the omni-present genetic shuffling, guarantee this primary rung. From all this available variability in the genetic material, natural selection will determine the phenotypic characteristics that will linger in the members of a population, as well as those that will be negatively selected according to the environmental conditions at the time. It has been documented that a wide polymorphic condition, different genotypes in any population, provides an ample response in front of possible environmental changes. In addition, intra-specific competition could be reduced with only small differences among phenotypes.

A reduction of polymorphic variants in any population implies a certain degree of specialization, which is unfavorable if sudden environmental changes occur, since not enough survival and reproductive responses will be available. The genetic mechanisms that provide and keep the polymorphic condition are exposed to selection, then becoming elements of adaptation; hence, it is rational to judge them as *adaptive polymorphism*. It is by no means untrue that, in very special cases, certain polymorphic variants may be neutral, but it is not a rule; polymorphic loci may act directly on certain adaptive traits. It is also important to consider that, as any other gene in the biosystem, polymorphic alleles are involved in gene interactions; considering polymorphic genes as isolated genetic entities is a more serious problem than grant them adaptive value to some extent. By all means they are an active part of the huge biological complex, where *genotype* and *population gene pool* are framed in the particular and in the general sense, respectively.

The great genetic variability in populations is considered the central problem of Population Genetics. The environmental heterogeneity could be a possible explanation to a certain degree. Besides, it has been pointed by different specialists that spatial ecological variation seems to be more important than time variation in keeping genetic polymorphism. This is consistent with the criteria regarding an ample gene pool for any population, to efficiently respond to new changing environmental conditions, at least, with a certain amount of genotypes far from the critical genetic limit; extinction is foreseeable, otherwise.

Variability and diversity, as evolutionary categories, are the ones that confer the general concept of biological evolution. It has been stated by an outstanding evolutionist that the best definition for the biological evolution is to consider it as a change in adaptation and diversity in organism populations. Likewise, other authors suggested that, at least a certain portion of the variation observed in populations could be correlated with abiotic variables of the environment, according to their results on natural population studies.

More than thirty years ago it was stated, as a condition, that a variable environmental setting for a particular species, would be assembled as a contradictory selective mosaic regime to keep a single locus polymorphism, based on a natural space environmental heterogeneity. It is understood that those selective regimes are contradictory since they are created from an ecological combination that changes along the space, producing an environmental discontinuity made of *ecological patches*, which in turn, build a true *ecological mosaic*.

In fact, the concept of ecological mosaic is considered important based on its evolutionary and adaptive connotation, which is clearly manifested. The basic element of the environmental heterogeneity, in its own space distribution, is the presence of ecological patches, within which uniform selective pressures do exist. The mosaic nature of the environmental heterogeneity is evident. Populations dwelling in these ecological patches show a trend toward the emergence of particular phenotypes, without gene flow, or a very low gene flow rate. Different authors have found these results in natural animal population studies.

Every time that we look at any phase of the evolutionary process, it seems like a tremendous, logical, and precise assembling is present, and not even the slightest detail is missing. On the contrary, the perfection is comparable to an incredible snap-on tinkering with the fitting of every part crafted to perfection. Nothing is misplaced, everything is included, and the process itself runs smoothly as if everything has been calculated, programmed, charted, and determined before hand. If all these aspects are necessary with such a perfection to accommodate and lead the process of life, it seems a doubtful possibility that the same perfection, timely developed and extremely delicate processes all advocating for the incommensurable surfacing of life itself, could be elsewhere out of our planet. Besides, such perfection elapsed in a period of thousands of million years, keeping the course, without drifting, at least regarding the main stream of the evolutionary process; it is too long a time for such a huge progression to succeed *without havoc* in the incredible itinerary, which gave birth to life, in so many different ways. It is striking to believe that such perfection as life is, just came from the chaos of primitive oceans that licked the shores in the evolving planet, millions and millions of years ago. And not only life, but intelligence as well, for we are not only highly developed animals, we have been touched with intelligence from which an incomparable quantity and quality of technology we all enjoy nowadays.

As we are used to see life around us every single day of our existence, we take for granted that life is a common issue, we do not even stop to think for a moment what an enlighten prodigy we get pleasure from, every day, as our own life elapses in our extremely beautiful planet, which is a "gift of nature..." Planet earth itself is an emporium of wonders in its immense journey of evolution. Our planet is a planet of water without which life could not be possible on it; at least, in the way and manner we understand life. Then again, every single element is present, nothing has been left behind. What a magnificent architect is "evolution" using its tool of wonder that "natural selection" is.

Another point of view may focus on the previous assertion: Is evolution the architect or the tool? And...is natural selection just the main outshine part of the tool? What the answers could be for the first and second questions? If evolution is the tool, who is using that tool? As a consequence, if natural selection is the outshine part of the tool...who manufactured that outstanding part? I do not dare to answer. I do not want to influence the reader with my very own opinions. This is not the objective of this book. The reader may use critical thinking...and arrive to his/her own conclusions and answers, to intellectually clarify the concepts that lie on the previous questions. This is by no means a doctrinal book.

Biological Structures and Variations

A discussion about genetic variation cannot be completed without the inclusion of the structures that conform and identify animals and humans; I want to refer specifically to vertebrates as the most advanced and developed beings in the animal kingdom. Many are the biological structures that show an extraordinary complexity. Each one of them deserves a separate discussion and consideration. All have relevant features as long as living creatures are concerned. I would like to discuss the issue following a deductive approach; it is, from the general to the specific level.

Any vertebrate organism (fish, amphibians, reptiles, birds, and mammals) is astonishing perfect. But what is most striking is the complexity of every single organ, which in turn conform systems, and all those biological systems work together in perfect coordination, till the pathology arises. Nevertheless, if we take a closer look to any organ or structure, we come immediately to an introspective reasoning, it is: I wonder how so many marvelous biological structures, their functions, and their incredible coordinated actions came to be exclusively following the rules of evolution, no matter how many millions of years went by. The same time could pass without any result, or just small and insignificant attempts of evolution, restricted to the simplest expressions of life.

Looking at a tiny ant and being aware of its extra small organs and systems working with remarkable precision, it evokes the notion of nano-technology in an organism evolved millions of years ago. Now, as the opposite of such minuscule being we may observe a rhinoceros or an elephant, with big organs also working efficiently and coordinated, not less than the ants' organs and systems do. Then again, the "culprit" is: the millions of years elapsed that marvelously can "create" anything, because with such an immense time-span "anything is possible", from an ant to a dinosaur. In vertebrates, the blood is there because they have vessels to carry it, but at the same time they have a heart to pump the blood along the vessels of different diameters, some are so thin as a single hair could be. Besides, the blood carries the oxygen to make breathing possible and also nutrients for the cells to be alive and reproduce and repair themselves, and so, and so...It is too complex, too perfect; too many tissues and structures in an incredible assembling including thousands of living things, each one with its own biological requirements, restrictions, and possibilities. This is just the general aspect of the problem, the whole biological system working and preserving life for the specific animal, humans included.

Besides the perfect coordination among all organs and systems, there is something else. In the evolution of the living matter we can observe the sequence, it is, the emergence of organs necessary for a specific function. Of course, genes are responsible to determine the position of each single organ and every part of the body; that is not the point, the point is how every organ emerged at the precise moment, at the precise place, for that organ was needed to complete the system. How could chemical or physical processes know about the organs needed? Take an example. The liver is essential to the digestive system, besides being a hematopoietic organ. What was there before the liver emerged? Was it another liver but imperfect? With an imperfect liver life is not possible. It is or it is not. Nothing is between. The liver is just an example but the same is true for the rest of the organs in any living creature. Are the purest evolutionists to conclude that little by little according to the needs of the evolving creature, the organs and then systems came to be? If it was little by little, no living creature may strive with lacking organs that proved to be essential for such a living organism. An earthworm is very different from a vertebrate; it does not need as many complex functions as vertebrates need. But vertebrates did not emerge from such annelids, hence we cannot compare higher living forms with simple ones just to say that the latter has or does not have this or that. That is out of the question, besides being pure speculation. I understand that the *only possible floating board* is: *millions and millions of years elapsed,* to explain what cannot be proved. Besides, the earthworm itself is not a project in progress; it is a complete and adapted living thing, by the moment time. What they have is exactly what they need. Is any zoologist or pure evolutionist around there involved in any project to improve earthworms or any other member of the *Kingdom Animalia?*

But there is something more striking and very common at the same time, which is built of different parts each one being a true unit and then ensemble together to structure a whole system. I am talking about the skeleton, whichever vertebrate animal you want to consider. In any skeleton every single bone is a true unit by itself and connected to the next unit (bone) by a joint, which could be flexible or rigid. The bones of the skull share rigid joints, the bones of the hands share flexible joints. Think about any skeleton, from a whale, from a wolf, from a cat, or our own skeleton. All of them are structured with single bones joined together to assemble the skeletal system. What was the first bone to appear? Perhaps the question is non-sense since a single bone, or two or three means nothing; it is a system, without its own structure isolated bones are meaningless. But the problem remains...how the skeleton came to be? How all these bones "evolved" to fulfill the task according to their specific placement in the body? There is no mistake; each bone has been located in the right place. I had never heard an orthopedic specialist complaining about the structure or function of any bone in a normal human being.

Above all, besides being a whole system, the skeleton is the support of the body, and the frame to allow movements that are essential to the high-developed living matter. But there is more based on the above. Muscles are connected to bones in a perfect way, in the exact point, not less not more than necessary. Then again, the purest evolutionists will claim that genes are in charge of placing every muscle in the right position and with the right connection for every single bone. That is not the point, everybody knows what embryology is, and genes are in fact responsible for the perfect ontogeny process in living organisms. The point is *how* bones and muscles came to be so marvelously hinged, to produce the final outcome of movement-support-frame to make possible all the tasks animals are able to perform in the dynamics of life. How the skull bones were assembled that the brain is cased inside? Was it really the result of natural selection through millions and millions and millions of years? Who was aware that the brain was necessary, and besides its specific location where it is, and nowhere else? Is it possible that evolution located the most primitive brains somewhere else? It seems to be impractical. The brain is where it is today since the most archaic vertebrates appeared on the stage of life. I will insist again, the point is not to argue if evolution did or did not, such and such. The most important point is the **sequence of events**, not the events themselves. In the sequence lies the mystery of the evolutionary issue, not in the characteristics or specific parts of any animal. Without the **correct sequence** at the **exact time** in the architecture of biology, I could not be here writing this book. Some specialists have stated that a certain structure may evolve to fulfill a specific function and then might be ready to serve and exert another function, so the *de novo* development is not necessary. It seems to me that such a statement means that certain organs are so lucky as to be ready to serve more than one function on a primordial state. Besides, it sounds like a multiple-function-

determinism on evolutionary terms. Furthermore, according to such postulates genetic changes seem to be *poly-functional oriented*, rather than single and isolated processes as mutations have proved to be. This theory has been named **exaptation**; again, a new trait comes up to serve a function long before employed in another. This exaptation could be an isolated phenomenon to explain certain occurrences, but this cannot be a rule to explain all the high complexities of all biological sequential events that lead to extraordinary living organisms.

If you want to construct a car, every single part has to be in the right place, at the right moment because you, the constructor, make it happens that way. A car is a very simple task compare to the living matter. Imagine how could natural selection that is simply represented by multiple selective forces of the environment, be engaged in such an immense task as that of creating life in millions of forms, present and past, with the only aid of millions and millions of years. If we focus the problem of evolution in that way, it sounds like a tall tale for adults.

It has been written that creationists, I am not one of them there is a big difference in between, see a design or a plan regarding the living world, while a scientist see an order or just a regular arrangement. This is an euphemism, since any order or regular arrangement is in principle a design, perhaps a rough design but it is. You cannot plan or design without an order. Order is a premise to obtain a logic design. Besides, evolutionary fundamentalists claim that order is not evidence of design. Again, the first step for a design is the order of the elements, and order itself is still difficult to be the product of blind forces of probabilistic nature, as is the complex relation among environmental elements and selective pressures, which are their manifestation. When I refer to order I mean the order present in living things, not the mere order of rocks or pebbles somewhere around a shore, because of the dynamics of tides.

What evolution explains is how modifications of different bones were accomplished giving birth to the skeletons of the five classes of vertebrates: fish, reptiles, amphibians, birds, and mammals. Besides, there are ostensible modifications in every class, which is very simple to realize. As an example, we may observe modifications (variations) of both the skeleton of a frog and the one from a salamander into the Class Amphibia. Additionally, apparent modifications exist among birds; a hawk's skeleton is not exactly like a parrot's. The same is true if we compare a rat's skeleton and the one from a cat. Likewise, other biological systems ought to be scrutinized under the same focus as has been done for bones and muscles; the same conclusions will be reached: the problem is not to say that genes are responsible for every single detail during the ontogeny process, it is well known and understood since many years now.

For the rest of the systems, including all the organs as peerless parts, the question is the same: how the system came to be in its highly complex configuration, and how the organs comprised in the system emerged to make possible the extremely coordinated biological functions without which life is not even a mere project? Life would not be possible either for the simplest organisms on earth, without the state of perfection of internal biological systems' coordination and arrangement.

Now the arthropods, those animals like insects, spiders, shrimps, and lobsters that have and external skeleton. Everybody knows how hard the external part of a lobster is, it is so just because that is the lobster's skeleton, known as exoskeleton, because it is placed covering the whole animal from outside. No matter if it is outside or inside like in the vertebrates, the point is that we are dealing with the same concept with two different designs: exoskeleton and endoskeleton. Both are adaptations, so far very different but making the same biological function, it is, to sustain the body and make the movements possible. Well, now the question is: did natural selection opt to split the skeleton concept in two different designs to comply with what? Those of shrimps and lobsters may be explained as protective skeleton structures since they are sea-dwellers. Well, how about mosquitoes, flies, bugs, spiders, scorpions…they are not sea-dwellers, they are terrestrial. The explanation has to be different. Then, according to the title of this chapter…why such a genetic variation? To sum up, two different skeletal systems were selected to fulfill the same biological function, which is not very efficient in terms of evolutionary processes.

Another biological system that deserves special attention is the nervous system. As it is well known, the nervous system comprises an intricate and complex organization of nerve cells (neurons) that in general terms connect all parts of the body with the brain. Let us refer only to the nervous system in vertebrate animals, though lower forms of life also share the same attribute in a lesser complex design. Nevertheless, how could it be possible that natural selection guided the construction of the nervous system, from very basic designs along through the perfection that we find in vertebrate animals? If we consider that, natural selection alone was able to guide the construction and final touches of such a magnificent system, in fact we are declaring that natural selection is an extraordinary combination of natural forces bordering intelligence.

Without a nervous system is not possible for most of the living world, I mean animals with complex biological organization, to perform all reactions connected with the most elemental activities of living organisms, such as reproduction, food search, survival reactions, seek for shelter, avoid dangers, communication, just to mention the most common ones. Besides, without a nervous system no movement of skeletal parts could be possible. As a consequence, without a nervous system animals would be just like living plants. Then again, how could we explain that the nervous system appeared to make possible animal life on our planet, from a simple ant to a dinosaur? How could natural selection know that a nervous system was necessary to create real animals, complex ones? Evolution could stop its task in living organisms without a nervous system, and then planet earth would be a planet of plants and of very basic animal forms, like protozoan. Who was aware that the presence of nervous systems in the living matter would make the difference between a plant and a complex animal? Was it natural selection?

It is hard to believe that all selective forces during no matter how many millions of years, could lead the tremendous engineering task of constructing nervous systems in animals, besides being aware that nervous systems make the difference between plants and higher animals, as was stated before. Then again, without that system this would be fundamentally a planet of plants, aquatic and terrestrial; then, humans could never be on earth. This is not the dim controversy of evolution or creation. **Evolution is an incontrovertible fact.** But there are many aspects of evolution that science cannot, at present, elucidate in a convincing manner. Many gaps are unsolved, not to mention the processes extremely complex to be plainly and ingenuously explained by environmental pressures and natural selection, acting upon phenotypes.

Most of the evolutionary fundamentalists' claim, as the only plausible or best convincing guessing, that millions and millions of years of unstoppable evolution can explain all the accomplishments the living world is showing today, and in the past. This means, according to their "time-related-expression" those million years can make miracles. In that way these fundamentalists are endowing time with attributes of a real God, instead of searching for better explanations not to repeat over and over again the same speech, claiming the improvable elapsed time as the *non plus ultra* explanation for what they are unable to peer in. All of them repeat the same, proving a lack of initiative or an evident impossibility for delivering better convincing explanations than this: *time is the only responsible for the magnificent expression of life on earth with the aid of natural selection.* They stick to the appearance of the phenomena, admitting nothing but variation-natural selection-adaptation and then evolution. Yes, all of these events are irrefutable, but they admit nothing else.

There are not only missing links in the fossil records, unfortunately, we are having missing links in the course of reasoning, since the focus, at following these true steps, is somehow distorted just because the **essence** of so many (millions) complex adaptations are overlooked in the name of science. They are reluctant, or afraid, to incur in somehow "unusual" considerations not accepted by the official status of nowadays science. They tightly stick to the most orthodox concepts; they behave like top of the line fundamentalists, evolutionary fundamentalists.

For many of the characteristics in the living world natural selection is not enough to explain the course of events, remember that natural selection is only *a tool* that acts upon phenotypes depending on many selective forces present in the environment, but natural selection aiding time is not enough to explain thousands and thousands of species living and extinct, most of them unknown to science. Many specialists have tried to speculate how many species have been extinct, but that is a mere guessing, there is no ground enough to really assert the amount of extinctions that account for the long trail of evolution on earth. Is natural selection and million of years so efficient and creative to push evolved so immense number of species?

One contrast that is needed and not available is the certainty of life in another planet, somewhere in the universe. If humans could ascertain life in any other planet, then a contrast, a real scientific comparison would be available, and many hypotheses could be tested, then accepted or rejected, in the same way zoologists compare a prospective new species to the ones most closely related and already known, to accurately determine the taxonomic characteristics of the new one, the ecological relations, the proximity or remoteness on a zoological basis, to finally determine what status is the proper one for the new form available. Unfortunately, such a possibility is not at hand for human evolution final contrast, and perhaps will not be ever.

Evolution...not a continuous variation process

The continuous variation postulated in On the Origin of Species could be not exactly that way. First of all, fossil records point to another interpretation, and in just few cases a continuum is evident. The principle mentioned by Darwin *natura non facit saltum* appears not to be so accurate. In fact, different biologists are theoretically prone not to accept as a final truth the unchangeable condition of slowly and continuous evolutionary change. The questionable issue arose when no fossil evidence was achieved, in spite of a huge amount of fossil records, showing intermediate forms or any proof of sequential development, leaving the paleontologists with the only acceptable tool of comparative anatomy for the purpose of classifying and projecting relationships among and between extinct species, and modern ones. Of course, isotope dating is a formidable technique in assuring the time-relation in the evolutionary context; but the appearance of intermediate forms is not the rule, albeit the amounts of available fossil records ever since. The only possible and logical question arose many years ago: are intermediate forms a reality or just a hypothesis that needs to be proved? All evidences pointed to the hypothesis criterion, hence it was mandatory to test it. The first step in such a testing procedure was to implement a consistent explanation based on the practical experience provided by the fossil records.

In the 1940's an evolutionary biologist, Goldschmidt, stated that drastic changes may occur in the evolution of species, probably by macromutations. Such a declaration rapidly encountered opposition by a great number of recognized biologists. It was Goldschmidt who for the first time established the difference between microevolution and macroevolution, and stated that an accumulation of micromutations could explain variations within a particular species, but never will produce a new species. In fact, Goldschmidt approached the concept of *punctuated equilibria*, which will be discussed in this chapter. The concept of macromutations stated by Goldschmidt implied isolation for the target population; without the isolating effects a new species could not be formed. Of course, the problem understood nowadays, is not the issue of macromutations exclusively. Little and apparently non-important mutations can lead to metabolic changes that affect other important traits, within a short geological time cause differentiation among or between groups, resulting in a new species with no intermediate forms.

The Darwinian theory of speciation states that, a gradual acquisition of novel modifications must lead to the emergence of a new species. Nevertheless, Darwin gave the possibility of rapid speciation for those groups in farther areas of the original species distribution. In modern times it was used another term: *phyletic gradualism.* Gould and Eldredge, opposing to phyletic gradualism, launched a different postulate in 1972, known as *punctuated equilibria,* already mentioned before. Briefly, punctuated equilibria establish that species endure long periods of no change, which might be as long as millions of years, then "suddenly" a new species appear in the geological column. Of course, the term *suddenly* refers to the geological time, not comparable to human lifespan. This theory was based on their paleontological finds, which pointed, according to the fossil records, to another explanation different from the phyletic gradualism; presumably, phyletic gradualism is rather rare.

The problem was that many biologists misinterpreted this new approach, and then a somehow distorted criterion spread upon the realm of evolutionists. Punctuated equilibria do not mean that transitional sequences do not happen. Neither is phyletic gradualism the mechanism of evolutionary change, nor the exclusive way of speciation; this is not in contradiction with the existence of transitional sequences. Furthermore, one important aspect of the theory of punctuated equilibria is concerned with the relative high frequency paleontologists' finds match the punctuated equilibria model compared to the phyletic gradualism enunciate. Hence, in fact, the theory of punctuated equilibria offers an explanation for the most frequent patterns paleontologists find in the geological column.

The fossil record is utterly incomplete; it is expected this way since the impossibility of fossilized remnants of all extinct species, is a logical fact. Nonetheless, the theory of punctuated equilibria offers the elucidation for the amazing sudden emergence of species, the records of extinction of species, the transitional traits in fossil records if available, and the relative steady morphology in those extinct species that enjoyed a widespread distribution.

One of the most common patterns found in paleontology studies is the so called *stasis of species* that may endure for long geological periods, and then suddenly changes occur giving birth to a new species, that in terms of geology abruptly appear all of a sudden, as stated before. Then again, most species behave this way according to fossil records all over the world. This repeated pattern of appearance cannot call for a mere stochastic process but a true evolutionary blueprint, which point to the fact that among other possibilities, for speciation and evolution as logical processes during the existence of the living matter on earth, sudden appearance in the realm of species is the magic touch for extant and extinct species as well. However, a full convincing hypothesis to explain why the sudden emergence of a species occurs as the most common pattern of speciation, is not available.

Punctuated equilibria only offer the theory to recognize that phyletic gradualism is not the exclusive mechanism for speciation, and with this theory paleontologists are provided with an *ad hoc* explanation to support their finds in the geological scenario. Perhaps the two most important points claim by punctuated equilibria establish that: a)usually species occupying extent territories show a very slow change, if any; b) those populations that give raise to new species usually are found in very small restricted areas, it is, their distribution is fairly limited, hence sibling species evolve in a limited geographical environment. It is very important to acknowledge that punctuated equilibria is not a theory of macromutations. Nevertheless, Gould acknowledged four kinds of macromutations mainly:

a)   completed adapted species arises from a single mutation.

b)   Swift reorganization of the genome due to reorganizing macromutations

c)   Not perfectly adapted macromutations arise; then partially adapted.

d)   Macromutations on genes controlling embryo development.

All of the above is a tight account of the prevailing theories that exist under open discussion at present. Such arena of moving ideas permits a subtle message to emerge from the sometimes confusing arguments and responses launched by eminent specialists. It is out of the question that as time goes by better approaches are in sight, trying to explain in an advancing parade what species evolution is. Take this into account; almost no specialist is much concerned with the evolution of superior taxa, simply because the key to answer the emergence of such a phenomenon as the evolution of species is also the key to answer the materialization of life, to some extent.

In fact, up to present it seems that many are the trails that have been used by evolution to create new species, according to the proposed theories and reasoning. Besides, it is really astonishing to think of the enormous figure that may represent the extinct species, those which no human eyes ever saw. What we have today as extant, is nothing compared to those that vanished along the incommensurable flowing of time. More theories are yet to come, no doubt about it, human intellectual capacity will endeavor in the future probably in a more efficient manner than we do today.

During recent years, strong evidences point to the fact that there are inherited variations very independent of variations in the DNA molecule, which means that they possess a not well-understood autonomy from DNA, to some extent. Actually, inheritance goes beyond DNA. Furthermore, it appears that is not the gene the target of selection, but rather a complex network of developmental interactions. In this context, the expression of the gene is not exclusively based on its own nature, but on a regulatory influence of such developmental network to which it belongs; multiple gene interactions and their expression are, apparently, the real target.

Two important concepts have been developed and studied recently, *epigenetics* and *epigenetics inheritance*. The first is connected to a complex of regulatory mechanisms, which may direct consistently developmental changes, such as variations in states of the cells, inherited through cell division. The second, is itself part of the first, includes social learning, interactions between the individual and the environment, and for some specialists the mother-offspring interactions. Then, we can talk of epigenetic variants. Unfortunately, these studies have been carried mainly on bacteria, fungi, some plants, and few invertebrates; even though, it seems that epigenetic inheritance is a real fact from one generation to the other, in every organism.

In addition, environmental pressures may provoke variations as an epigenetic inherited trait. The central problem in all these cases is that they sound like a Lamarckian proposal (Jean-Baptiste Lamark, French biologist 1744-1829, who put forward the concept of "inheritance of acquired characteristics"), which prevented them to be accepted among biologists. Nonetheless, many evidences are at hand that denote the fact that certain chemical compounds as well as other environmental factors, may alter the action of genes.

Other terms have been brought to the specialized and theoretical literature. Perhaps, one of the most difficult to identify and make a definition of is "semantic information." Some scholars recognize how difficult is to accurately define such a term. Basically, it is connected to, or comprises, all elements of selective environments that are recognized as genuine, regarding the fitness of life forms. The so-called semantic information only has a complete meaning if consider in the context of particular selection pressures on specific organisms living in specific environments. The definition includes DNA, RNA, as well as other molecules. It seems an acceptable theoretical approach as intent to explain, al least in a coherent frame, the fundamental elements involved in a facet of the evolutionary process. Nevertheless, *semantic* means the branch of linguistic science, which deals with the meaning of words and the change of these meanings. We are talking about biological phenomena; I do not see the reason to link semantic with evolutionary definitions, besides semantics has nothing to do with the evolutionary theory whatsoever.

This definition could be important from a theoretical point of view, since we ignore, at present, the true order of appearance of "semantic information" (if it is etymologically correct), environmental resources, metabolism, and DNA replication. Precisely, this is the central problem to understand the origin of life, what was the order in which these elements appeared in the evolutionary context? These new ideas and concepts, unfortunately, are waiting for more information, since the available one is only the doorstep of a possible new open field for understanding human evolution. For this purpose, I have included these new theories, given that, in a near future, they could be of a great importance to better approach *Homo sapiens* evolution, and the reasons for this species to be. At present, it is only theory with some interesting evidences in certain low zoological groups, but a little distant from any near approach to human evolution.

In fact, nowadays, we have come to a detail and deep knowledge of genotypic variation according to some specialists. Genome studies abound, and laboratories techniques have advanced to great lengths. On the other hand, we are not in good shape of knowledge regarding phenotypic variation, according to the same specialists. Perhaps, this is the main reason why evolutionary processes have not been well understood; it is a drawback for scholars who attempt to progress in evolutionary theory.

Some authorities are prone to set apart phenotypic variation from genetic variation. Genotypic and phenotypic phenomena both are the sides of the same coin; the general concept of genetic variation may include both genotypic and phenotypic variations as well, since there is no possible existence for a phenotype if it is not genetically based. Definitely, the phenotype is a major expression of the genotype, *sine qua non* the former cannot exist, albeit it is not one hundred percent dependent of the genotype, since environment plays an important part as well, together with other cell interactions, on a deterministic way in tissues and organs adaptations.

Then again, if we talk of a theory of evolution, I would rather take, as main components, genetic variation, mode of inheritance, and natural selection. It is undisputable that phenotypic variation, if you prefer, is in all a complex outcome. Nevertheless, it is a logical approach to start with genetic variation and then on. Besides, it has been clearly exposed by some specialists that, genetic change and mutations are the leading elements for any kind of regulatory change in the organism. Above all, mutations are nothing but a very specific kind of genetic change at the molecular level, then, if we use genetic change and mutations as separate concepts, this will sound like a runabout circle.

Another controversial concept is *evolvability*, which, in simple terms is nothing but the capability of species to evolve. The central concern with this concept is that it postulates: complex life forms can emerge simply based on selection pressure acting upon random genetic variations. Philosophically, it confers a random origin to life, in all its forms, without taking into consideration any other plausible explanation, but randomness. It is, the high possibility to evolve based on natural selection. On the other hand, the amount of evidences is not enough to attest for the random-based ability to evolve. At present, theoreticians have to be careful enough not to create a confusion of terms in their intention to classify and dissect the elements of the evolutionary theory. I applause the theories, but they need to be proved beyond any logical doubt. Scientists have to take precautionary steps not to launch their theories beyond the reality of scarce evidences. The central problem of evolvability is how to effectively measure it. Besides, scholars have openly declared that the true factors affecting evolvability and its own evolution are weakly understood. Taking into account all these topics, it appears that they are the foreground of a non-completed evidence of proved facts; there is a long way ahead to be covered yet.

The Species Onset

Fossil records will increase in number, but what is difficult to accept is the fact that species come into view, establish a residence somewhere on our planet, decrease in number and…finally disappear. What is the purpose of such difficult, painful, struggling, titanic development to get into nothing…at best a few fossil remains just to let us know that there was a time, in the incalculable past, that such species existed? To the most inquisitive humans, questions arise about the existence of those species that are no more; how really were they in their archaic environment? What ecological relationship did really exist concerning those populations? How hard was living for them? How extinction came to be for every single species that we never saw alive? If extinctions were the direct result of drastic environmental changes…how stabilization for thousands of species could be possible in such a way that they got along through those destroying environments up to present as evolved descendant species?

Some 534 million years ago, the first primitive metazoan began to emerge during the well-known Cambrian explosion. From there on, body plans of metazoan began an evolutionary process, and not less than twenty million years ago all triplo-blastic (having the body with three tissue layers: ectoderm, mesoderm and endoderm) metazoan body plans, evolved. The metazoan radiation exhibit a very important genetic aspect, regarding a cluster of common extremely conserved gene products, which has been observed to generate animal body plans, since the beginning of this group. One intriguing aspect is the fact that, how was it possible for metazoan complex structure to be achieved with an unchanging group of elements? To create tissues, as metazoans possess, it is essential the appearance of new genes with their corresponding products to accomplish such a function. Remember that, the acquisition of new functions for already established genes is the so-called exaptation, mentioned before in this chapter. Probably a similar process was responsible for the creation of tissues, it is, the aggregate of cells with specific functions.

Facing so many *incognitas* and so few answers it is frustrating to conclude that life on earth was a fortuity, then, is the law of the universe an appearing and disappearing event? This cannot be the approach to such a complex development of living organisms. Furthermore, if life is not an exclusivity of our beloved planet, and other planets enjoy the same possibility of having plants and animals...is the universe so a simple system that after endowing life on planets other laws destroy them for good? Nothing in the universe is just for the fun of it. Nothing in the Universe is just a mere coincidence or a stochastic process. The problem is ours, not to understand what the phenomenon is. Remember Galileo, the Inquisition tried to silence the truth, and in 1633 he was forced by the Inquisition to retract. It was not until 1992 the Church granted recognition, almost 360 years later. The arrival of proofs and the smashing forces of evidence urged the belief systems to shut up and gave way to the scientific truth, in spite of the official preaching. Nowadays, not even the most stupid and dumb human being dare to dissent on the revolving movement of planet earth.

At this point, it is mandatory to establish a definition that perhaps has not been stated clearly along this and the previous chapter. It is not the species as a whole the biological complex, which is susceptible as the unit for evolution, it is the population the elemental evolutionary unit, though natural selection acts upon individuals they are not the target for evolutionary processes either, since individuals do not carry their own evolutionary performance, simply because of their extremely short lifespan.

Of course, individuals are the construction units of populations but not the population itself. It is worth to say that the genetic endowment of a population is nothing but the addition of the whole complexity of genotypes of hundreds or thousands individuals comprising it, sometimes millions of individuals. This complex sum of genotypes is the *gene pool* of any population, and it is this gene pool which is the target for the evolutionary processes to act upon.

Obviously, evolutionary development (known as Evo-Devo) starts into the population complexity. If the population has enough genetic variability in its gene pool to respond to environmental changes, no matter how drastic they might be, adaptations will surface and the population, besides augmenting in numbers of individuals, will show an evolutionary progress in terms of better adaptations; we cannot, or we may not, talk about evolution, otherwise. In fact, a true complex gear of developmental renovation takes place between genotype and phenotype, and precisely, is the novel approach of Evo-Devo that has started to clarify the evolutionary role of this exceptional gear. Another similarly important concept is the emergence of phenotypic novelties and their evolution. It is of the major importance to distinguish between novelties and common phenotypic variations. This differentiation between both concepts will open a new discussion and approach to a modern view on the evolutionary theory, as an addition to the Modern Synthesis.

According to Evo-Devo, innovation and phenotypic variation cannot be treated as alike. In a previous paragraph, the Cambrian explosion was mentioned. Then again, it is believed that many of the body plans that emerged during that period, the metazoan anatomical features, were not the direct product of genes but the combined physical properties of cells, helped by gene products to some extent, but not due to exaptation. Nevertheless, a group of specialists sustain the concept that genetic evolution leads to phenotypic evolution and novelties.

Not all populations of a species are able to turn into a new species. Additionally, the own concept of *species* is controversial. From 1940 through 1995 thirteen different concepts of species have been published and discussed in detail, which demonstrate that many theoretical approaches are possible. Recently, de Queiroz, from the Smithsonian Institution, has proposed a smart unified species concept, which endeavors to reconcile all the former concepts in one logical and functional perception of the species. The speciation process depends upon many factors, which can be briefly condensed in two major factors: genetic endowment and environment, being each one of a tremendous complexity, hence the complex interactions between genotype-environment that take place constantly in any wild life population; humans, albeit not living in the wild, are subject to environmental pressures, too.

Mutations begin affecting populations, covering gradually the distributional area of a species, and every single step on their course, affect a particular population. The larger the population the better the gene pool, in terms of genetic variability, which is substantially adapted to changing conditions in the typical environment of the species.

The isolation of a population preventing the gene flow or free interbreeding with other neighbor populations of the species, might be the first step towards evolutionary processes to take place. If the population is small, a certain degree of inbreeding will take place, and then the stabilization of prevailing genotypes will be an expected outcome. After many generations, the number cannot be precisely determined beforehand, it is possible that those genotypes may differ substantially from those of the original isolated population, and then the new gene pool would not be the same either. At this point, it depends on how different those new prevailing genotypes are compared to those of the other neighbor populations; if the difference is rather substantial, chances are that a real evolutionary process toward a new species could be taking place. Whether new species will endure or not, will depend on other kind of complex conditions of multiple factors, again, genetic and environmental factors. It is, though isolation could be a first theoretical step on the way to a new species, it does not mean that isolation can guarantee the success of the potential species. Isolation is just a condition of environmental order as a mutation and crossing-over are phenomena of genetic order.

Neither the environment nor the genetic endowment may guarantee the final success of any species, what guarantee such an evolutionary course is precisely a positive interaction between the prevailing genotypes in the population (gene pool) and the environment. If environmental conditions may change, and the established population lacks the adequate genetic richness, the survival prognosis for that population is poor.

Another aspect that cannot be overlooked is the fact that though species may linger for thousands or million years, all species...let us say almost all, are subject to a tremendous effort to sustain life. You may get a hold into the natural history of any living creature on earth to be acquainted of this assertion. If you really go deeply inside you will find how hard, many times thorny survival is, including the "apparent" success this or that species enjoys. Examples abound; just to mention one, the emperor penguin. These birds have to travel by walking more than 70 miles to mate, step by step with their slow marching pace, like an army in a tight row to protect themselves. They travel from different places, and all groups arrive almost the same day, sometimes almost at the same time. They simply take the chances of such a journey because the ice is solid and secure in that selected mating place. They get together by the thousands, and usually it takes more than a week to find a mate. The female, after laying one single egg, give it to the male, and then depart quickly to feed miles away in the ocean; she will die of starvation, otherwise. Another hard journey, this time to reload food, then back again to feed the chick with the nutrients she engulfed, since it was born in her absent under the male extreme care and vigil. Some females never return, the hard journey is a path to death because of cold, starvation or both. The chick will not survive either. The male shall care for the egg for 125 days; it is the incubation period for the emperor penguin during which the male will not eat at all, he will only take care of the egg and will stand firmly the blizzards, which are a common issue in those extremely cold regions of the planet. If the female arrives in time the chick will survive. There are chicks that die because of no returning mothers, others die because cannot stand the blizzards though the father takes extreme care to protect them.

The cycle will start again next season with no guarantee that next time will be better. This is not the only example of harshness in a species life cycle. If the reader is really interested to get a full knowledge on any species in particular, search carefully for the natural history of the species. You will be surprised. To any human being a reasonable question may arise: Why so inflexible? Was this the best way natural selection found to make emperor penguins survive? Was there another less harsh road to go along, and survive, and be a successful species? Is it a punishment or a fee we have to pay to be on earth?

What is a rational explanation for all this suffering from simple animals to humans? We are not excluded, of course. Do animals have also to pay for any sort of original sin and their suffering is just a consequence of it?

Life has never been a simple path. Neither has been for our hominid ancestors nor for us; only, exceptional cases, can tell their life-stories other way. On the average, humans suffer much more than any other animal. First, just because we possess consciousness, which means that we do not only stand material suffering but psychological and emotional suffering as well. If any close relative of ours is going through difficult moments regarding a threatening-life disease, we suffer together, for we are conscious of the whole consequences regarding the health problem; animals do not. If we are the one diagnosed with a fatal or bad disease, we again suffer emotional and psychological stress, worsening the process because we can mentally project the consequences in the future, in short and long term views. If those diseases strike our children or spouse, so much the worse; probably we shall not be the same for the rest of our lives. Animals never ever suffer in such a way. Someone may claim that dogs suffer when their masters are gone, it is true, but they do not live a conscious suffering as we do. They would not be in despair for the loss of their masters. We may be in dark despair for the loss of a loved one. Moreover, millions of humans suffer from famine all around the world, helpless, without a hope, unable to get basic foods and shelter, no matter how hard they try, and all this happens for many different reasons, due to social and cultural structures of modern life on earth; this situation is real all over the world.

Compare to other animal species humans, in many cases, are unable to solve such situations by themselves. Other animal species move to another feeding grounds or migrate to the proper place to survive; usually they thrive. Many humans cannot even migrate since they live bearing an extreme poverty, preventing them to make a move; no place to go, no way to move. Finally, many starve or succumb by diseases and epidemics. Migration for humans is not a simple task; it was rather simple for our ancestors when moving out of Africa, no borders and no visa requirements. Searching for food is more difficult. Animals forage freely; no apparent restrictions are imposed to free foraging animal species, but ecological systems. Humans cannot go out for free foraging as animals can do.

We are the rational species; we need codes for living in modern societies, logical restrictions of well-developed cultures. Animals are not rational, they do not need such social restrictions, at least not in the way humans need them. All this suffering, shortages, and poverty are not exactly what the rest of the species on earth suffer. They face such problems without thinking, they are not conscious what they are facing, and they cannot foresee the consequences or the future.

This is the price we are paying for being rational...and to sport consciousness; we are paying a high price to be humans. Natural selection and then evolution gave us a gift, the human condition, but that is not for free, we have to pay the cost of our own evolution. What is the meaning of all this? For humans, a short existence with emotional and psychological harshness is the rule; many carrying extreme poverty...and finally death. At least, animals are not conscious of those conditions, they simply live and act according to "impulses" of their own species, firmly conditioned through natural selection. If they get sick there is no way for them to realize what the consequences might be, they just keep on as much as they can. They bear physical suffering. We bear both physical and emotional, just because we are rational beings; the matchless rational being on earth.

Furthermore, planet earth is beautiful, but in this heavenly planet, we humans, suffer uncountable sadness, illnesses, and all kinds of misery. Is this the price we have to pay to live in such a wonderful planet? A planet full of wonderful places and landscapes cannot be enjoyed at full just because human beings are destined to suffer in many ways. Where is the human that can express as saying the following? "I live a wonderful life, I have no problems, I suffer of no illness, and my family is completely happy and healthy." I think those cases are really scarce.

From all of these discussed aspects on species evolution, it is more than clear that a final logical assertion leads us to recognize that the evolutionary phenomenon is not possible to be utterly understood apart from a depurated genetic knowledge, among other considerations. Ultimately, genetic knowledge means to look deeply inside the basic genetic structure, DNA, which is the starting point for any explanation concerning or connected to the enormous phenomenon of life on earth. From this (DNA) on, we find a formidable sequence of events that rush into the evolutionary context, pointing to different ways to accomplish a flourishing process, crisis of populations, instability of species, suffering for existence, and finally the inevitable extinction of every species, in a short or in a long term basis. All species will succumb in one way or another; none will endure... none has ever been. They can survive for millions of years as dinosaurs did, but finally they were swept off and remains scattered all over the earth are the only proof that such creatures existed in the gone past, perhaps for humans to realize the fantastic phenomenon of evolution of giants.

Dinosaurs roamed the earth from the Triassic to the end of the Cretaceous, approximately 150 million years comprising hundreds of species, probably thousands; paleontological studies cannot determine based on the fossils available. Modern humans (*Homo sapiens*) have been on earth for about 195,000 years according to the fossils found in 2003 in Ethiopia; compare to dinosaurs it is approximately 1% of the time dinosaurs lived on earth. Dinosaurs were not the first giants on earth. Before dinosaurs other gigantic animals roamed the earth, conceived as true monsters of the evolutionary process during the Permian period, and they also became extinct despite their powerful biological structures; life on earth became unbearable for them, and then for dinosaurs.

Very well documented studies proved that several extinctions have occurred on a periodical basis of about 25 million years on average, from the Permian to present, covering an approximate length of 255 million years. The two most catastrophic extinctions occurred, first at end of the Permian period, which according to different studies annihilated more than 90% of animal life on earth, including the vanishing of 60% of marine species; only about 4% of sea life made it. Trees on earth were nearly destroyed, drastically changing the landscape of our planet. The Permian extinction pushed life on our planet, to the brink of extinction. The second, approximately between the end of the Cretaceous and the beginning of the Tertiary period, some 65 million years ago, swept-off dinosaurs from earth. It has been determined an asteroid of huge dimensions impacted the Earth. About 50% of marine life vanished. Open scientific discussions to explain such extinctions have been sustained for many years to present. Different hypotheses have been on the scientific dominion, to explain and assert the consequences derived from the impact of the asteroid. During these massive extinctions not only terrestrial animals were involved, as it was stated before, marine animals had been as well. In fact, extinctions occurred but planet earth continued an ever-changing evolutionary course. Life prevailed on earth; extinctions, in the end, produced an incommensurable leap in evolutionary terms, difficult to understand but physically proved. Synapsids, mammal-like reptiles, which for more than 60 million years rambled on earth, were the prevailing vertebrates of the Permian. They used the same ecological niches as dinosaurs did millions of years after them. They were gone during the Permian extinction.

The landscape scenario of our planet today is very different from the panorama early hominids "enjoyed" in the farthest past of our history as a species. An ever-happening replacement of species is the constant factor we learned from paleontology, paleoanthropology, and paleoecology as new studies developed in our time have proved. A present observable example are crocodiles, they outlived dinosaurs. They have not changed during the last 70 million years, almost; the most powerful fresh water predator on the planet. Nonetheless, this is an endangered species, which populations are decreasing in many places nowadays.

They are entering the extinction phase in a slowly pace, but they are. They have to be protected; they will be extinct, otherwise.

It is accepted that modern crocodiles are survivors from the age of dinosaurs since they have been on earth for about 80,000,000 years. Today, only few species of crocodiles are found. The old ancestors of crocodilians, known as Crurotarsans, spread during the Triasic period in different forms all over the Earth, to become a dominant predator group. So powerful they were that dinosaurs could not compete for ecological resources. It was not until Crurotarsans disappeared that dinosaurs could take over.

Ecosystems change, and with them the different communities of species, comprising all imaginable combinations, giving rise to new species while others will disappear from the surface of Earth...and extinctions are forever. Most of nowadays species of vertebrates could not be here if dinosaurs would be alive. For us to be here, it was necessary the extinctions of dinosaurs as well as other predators, that could be a life-threatening enemy for our most primitive forerunners. In that way, certain extinctions make possible the appearance of new species, hence, the evolution continues, apparently without a directed cause, spreading its effects accordingly...things happen in a certain way, and others not expectable will take over in the intricate, unpredictable, mysterious, and not well understood evolution of our planet and of species.

In fact, species evolution is a very complex process, that is why so many alternate hypothesis, scientific reasoning, and... why not?...speculations surfaced trying to shed some light on the entwined scenario of the evolutionary course. Perhaps a single piece of daring logic may solve, at least in part, the somehow unreadable puzzle of the emergence of new species.

The main problem for this daring logic line of thinking is the impossibility of immediate testing, as any hypothesis must require. That was exactly the pitiful situation of Galileo, he could not test his hypothesis, and he was condemned. Science, in its rigorous procedures does not accept anything out of the book, out of the dogma. No scientist dare to face strong critics based on his/her deviation from methodology, or scientific methods. For me, it is exactly the point where the theory of evolution of species is at present. Just make a simple math: in 1859 *On the Origin of Species* was published, more than 150 years now still controversial points are on the stage regarding the evolution of species. Darwin has been strongly criticized, though his theory of evolution is the most outstanding and enlightened work on biological science, ever since. Anthropologists claim today that no explanation was given by Darwin's natural selection to understand what urged our pre-hominid ancestors to go down from the trees, and then became a bipedal species.

Why is so difficult to arrive to the right conclusion? How many generations of specialists have been on the business ever since and nothing completely clear is at hand? Probably because the scientific dogma forbids focusing on different alternate hypothesis, since those hypothesis are not orthodox and cannot be claimed as purely scientific, according to the official establishment. Let us time goes by, but chances are that we will not be on Earth (we will be extinct as individuals) when in the future some awesome scientist will dare to defiance the official status of science, and set forth what a myriad of forerunner evolutionists did not dare to.

Recommended Bibliography

Arthur, W. 2001. Developmental Drive: an important determinant of the direction of phenotypic evolution. Evolution and Development (3): 271-278.

Bates, M. 1990. Patterns in Evolution. Scientific American Library. N. Y. 246pp

Belyaev, D. K., Ruvinsky, A. O., Trut, L. N. (1981). Inherited activation-inactivation of the star gene in foxes: its bearing on the problem of domestication. J. of Heredity (72): 267-274.

Conway-Morris, S. 2006. Darwin's dilemma: The realities of the Cambrian "explosion". Philosophical Transactions of the Royal Society of London. B361: 1069-1083.

de Chardin, T.1965. The Phenomenon of Man. Harper & Row, Publishers, Inc. N.Y. 320pp

de Queiroz, K. 2005. A Unified Concept of Species and Its Consequences for the Future of Taxonomy. Proceedings of the California Acad. of Science. Vol.56 Suppl. 1. 18:196-215

Endler, J.A. 1977. Geographic Variation, Speciation and Clines. Princeton Univ. Press. 246pp

Eldredge, N. 1989. The Evolution of Punctuated Equilibria. Time Frames Series. Princeton Univ. Press N.J. 240pp

Futuyama, D.J.1983. Science on Trial. The Case for Evolution. 1st edition. Pantheon Books. N. Y. 251pp

Gould, S.J., Eldredge, N. 1977. Punctuated Equilibria: the tempo and mode of evolution reconsidered. Paleobiology (3): 115-151

Jablonka, E., Lamb, M. J. 2010. Transgenerational Epigenetic Inheritance. In: Evolution-The Extended Synthesis. Pigliucci, M, and Muller, G. B. editors. The MIT Press. Cambridge, Massachusetts. 137-174.

Fry, I. 2000. The Emergence of Life on Earth: A Historical and Scientific Overview. Rutgers Univ. Press. Brunswick, N.J. 344 pp.

Gould, S.J., 2002. The Structure of Evolution Theory. Cambridge, Mass. Belknap Press. of Harvard Univ. Press. 1464pp

Kirschner, M. W., Gerhart, J. C. 2010. Facilitated Variation. In: Evolution-The Extended Synthesis. The MIT Press. Cambridge, Massachusetts. 253-280.

Mayr, E. 1974. Populations, Species, and Evolution. Harvard Univ. Press. Cambridge, Mass. 453pp

Mayr, E. 1982. Systematic and the Origin of Species. Columbia Univ. Press. N.Y.334pp

Odling-Smee, J. 2010. Niche Inheritance. In: Evolution-The Extended Synthesis. The MIT Press. Cambridge, Massachusetts. 175-207.

Pigliucci, M. 2008. Is evolvability evolvable? Nature Rev. Genetics. 9: 75-82.

Schlosser, G., Wagner, G.P. 2004. Modularity in Development and Evolution. Chicago Univ. Press. 600pp

Strickberger, M.W. Evolution. 3rd edition. 2005. Jones and Bartlett Publishers. 672pp

Chapter 3
The Advent of Humans: the inexplicable trail

Human evolution began around six million years ago, when some hominid or hominid-like form split off from the common ancestor they shared with chimpanzees. About 2 to 2.5 million years ago some of those hominids left Africa. Unfortunately, certain fossils may indicate this first advance was not completely successful, though it seems they were at first. For anthropologists, it is not very clear why this exactly happened that way. Was it because the group or groups were too small and they failed to develop skills to protect themselves, to hunt together, to share a common living, to procreate in those harsh conditions, or too long a journey to stand on? Perhaps it was just a matter of bearing very primitive features to succeed in such an attempt where certain skills were needed. Besides, the reason why those misfortune groups decided to go ahead is by far less clear than the fact of their failure. They did not know anything about the ending of an Ice Age, for them to make up their minds to flee out of Africa, heading Europe and the Levant.

Moreover, Africa is a huge territory with plenty of possibilities to forage and find shelter; previous hominids lingered for thousands of years and never left. Fossil records do not account for such an explanation and in part, smart guessing is the only alternative we have at present. Geneticists have pointed that according to their research the most plausible event was only one single group of modern humans leaving Africa. It is just a possibility, if any former group was not lucky enough to succeed and then disappeared, or simply, no fossils have been found from previous departing groups, genetic analysis ought to be based only and exclusively on the available material. This is not the case for preceding *Homo* species as *Homo erectus, Homo heidelbergensis,* and *Homo neanderthalensis* (Neanderthals), which will be referred to in full details in this chapter.

It has been calculated that around 133,000 years ago a glacial period arrived to its coldest peak. As a consequence not only temperatures descended several degrees, a drier climate was installed as well. Dense forest turned into dry woodlands. Food resources changed dramatically leading to a fragmented condition for all populations that were seeking resources to survive while adapting to those harsh new conditions. It is not difficult to conclude that population dispersals could occur to some extent, leading to a diversification of communication skills, and probably modifications connected to cultural distinctiveness, turning out to separate tribes and cultural groups. Today, some authorities accept that probably several, *out of Africa,* migrating groups was the event that started the episodes of human evolution in Europe and Asia. It has not been consistently determined how many groups or what the intervals of migration were, but certainly that occurred. Recently, two scientific articles (2005, 2006) stated three different expansions according to results based on genetic markers. These three out of Africa expansions are timely coincident with the appearance of *Homo erectus, Homo heidelbergensis* and finally with the geographical expansion of modern humans, demonstrating an interesting and coincidental anthropo-genetic outcome. It is believed that *Homo erectus* was the first hominid to leave Africa, around 2,000,000 years ago probably trudging all the way to Southeast Asia.

We cannot expect fossil records may show all possible proofs to ascertain group numbers or migration events, neither rational tracing of real pathways used by our ancestors in their exodus when leaving Africa. Most of these clues probably will remain unknown, and just now and then sneaking elements of the real past will be available for researchers.

Most considerations on the migration issue are just smart guessing, not more than logical deductions at the light of scarce paleoanthropological and archeological materials, including molecular studies as well. The sciences devoted to study and disclose our most remote past must face innumerable *incognitas,* hard to answer, not only because gaps are abundant, but mainly because we almost ignore the real ecological conditions at the time. One thing is to approach the reality of that obscure past based on geological studies and another very different is the reality, itself.

In this chapter an extra effort has been made to cover the most important details, to my best knowledge, regarding human evolution, just to make clear for the reader how the road for our species evolution on earth has been so far, and mainly "to discover" what details are not clear, or are not consistent with evolution by natural selection.

Nevertheless, the reader should arrive to his/her own conclusions. The chapter is just information; the reader's intellectual ability is more important than the information itself. By writing, I **do not** intend to influence the independent criterion of any reader, I only and manifestly expose my points of view as a biologist, as well as my concerns, to share with all those that decided to read this book.

Perhaps the major and striking fact in our evolutionary history is the evidence of different human species (genus *Homo*) that lived in the past, some of them coexisted on earth, in different regions, and probably got in contact during an undetermined period not well understood through fossil records. Just to mention a few of those species, as the study of fossils permits us to plunge into the sometimes-undecipherable history of ancient humans, we may reckon the following: *Homo rudolfensis, Homo habilis, Homo ergaster, Homo erectus, Homo heidelbergensis, Homo neanderthalensis, Homo floresiensis,* and *Homo sapiens* as identified human species. Most scholars came to the conclusion that *Homo ergaster* was the real ancestor of all later *Homo* species, but at the same time its origin still remains as an enigma. It is too audacious to proclaim *Homo habilis* as its ancestor. Recently, in 2007, remains of *Homo habilis* were dug up in Kenya, which led to the fact that *Homo habilis* and *Homo ergaster* coexisted in our planet for approximately 500,000 years. Then again, the emergence of *Homo ergaster* is one of the most enigmatic outcomes in modern paleoanthropology. As an outstanding fact to add to *Homo ergaster* recoveries, I may mention the well-known "Turkana Boy" skeleton, found in Kenya, an adolescent that according to anthropological projections would have been as tall as almost two meters. In general terms, *Homo ergaster* specimens look more modern than those of *Homo rudolfensis* and *Homo habilis do.* Remains from these two latter species show a combination of fairly modern attributes and primitive ones, which is compatible with early *Homo* populations. Besides, all surmises point to the possibility that *Homo ergaster,* who emerged in Africa some two million years ago, was the true ancestor of *Homo erectus.* One important aspect to mention, which should not be overlooked, is the fact that in the last twenty years finds of *Homo ergaster* suggest it was a more variable species than formerly considered.

The ultimately unsolved mystery in the incomplete records of human evolution is the fact, I better say, the question...Why are we here and the rest of *Homo* species disappeared? It is not my question; it is the question of all scientists that, in one way or another, have been involved in human evolution and variation studies. Nonetheless, though the paleoanthropological evidence is unquestionable, no official or orthodox belief system mention the fact of different *Homo* species (human species) living on earth, linked to our remote past. For an unknown reason, at least for me, they try to establish *Homo sapiens* as the only, exclusive, and unique human being ever on earth. I do not understand either how such an unquestionable truth, as the finds of different human species remains, can be overlooked by belief systems.

As it was stated in a previous chapter *Homo sapiens'* oldest fossil was dated as far as 195,000 years old, from the Omo River in Ethiopia. On the other hand, the first record for the genus *Homo* was accurately dated as old as 2,350,000 years to present. It counts for a long journey, with very fast evolutionary results in terms of unprecedented peerless acquisitions, leading all this to ascending top qualities, which no other species on earth ever had before. To my best understanding, if it is necessary to enunciate what features make us really different from the rest of the animals, I may say with no hesitation that they are together: unrestricted bipedalism, language, and a rational mind; the latter as the superb characteristic that deserve a special rank, for making us humans. And these characteristics are exactly and completely unique to exist altogether in just one single species, *Homo sapiens.* Bipedalism is important, of course it is, but it is not exclusive human; penguins are birds and they walk on two legs, nothing extraordinary, but in penguins, language is not developed beyond any other bird species, using sounds to alert about dangers, to attract mates, and to protect their chicks. Many birds do the same. On the other hand, modern primates use a series of sounds, more sophisticated than birds do. Wolves, dogs, lions, and many other mammals communicate with simple or rather sophisticated sounds, which guarantee survival, pack cohesion, and reproductive success, among other objectives. What is an extraordinary event in human language is not just the fact of complete communication that led to human communities and social life. The most incredible event is the coordinated evolution of all the elements as vocal chords, tongue, muscles to move the lips appropriately, the pharynges, and the uvula plus the development of all neurons necessary to accomplish the act of speaking, and all these super specialized anatomical elements were present to evolve at the exact moment, in a synchronized fashion to start the evolutionary history of our precious capacity of language. That is really amazing. What an incredible coordination of evolution and what an extraordinary selection to end up with talking humans! If we rationally look at all these facts together, it seems like evolution and natural selection act as intelligent essentials to deliver a tremendous finished product: human beings on earth.

The discovery of the gene FOXP2, associated with the capacity of language has been a breakthrough in understanding how language was acquired by *Homo* species. I say *Homo* species, for this gene has no evidences of improving changes, at least important ones, in chimps. Of course, language ability came out only in humans, and changes in *Homo sapiens* FOXP2 gene account for those abilities.

According to researches accomplished this gene in humans manufactures a protein, which differs in two units from that of chimps'. Other genes may account for other differences between modern humans and chimpanzees. The chimpanzee genome project may provide enough information as to determine why they are apes and we are not. I may comment here something that, though trivial in appearance, could be essential if a true chimpanzee genome ought to be determined. As far as the information on that project is concerned, the only chimp species included has been the common chimpanzee (*Pan troglodytes*) and not the Bonobo (*Pan paniscus*). In a future chapter a broad discussion on this latter species will be offered, with the purpose for the reader to recognize the biological and evolutionary importance of the Bonobo in the parallel evolution with humans. Though some specialists may judge that the Common Chimpanzee and the Bonobo's genome may not differ significantly as a general principle, it may only be an *a priori* consideration, and results may prove very different, mainly if compare to humans. Unfortunately, modern science frequently fell short on important issues as it is. Other significant event neglected by science, and also related to chimps, will be discussed in that chapter, as one more proof that science ignores or undervalue certain occurrences that could be unique to shed some light on matters that have been set as incognitas for decades.

Getting back to humans, the exclusive, the real unprecedented attribute is our rational mind. Probably, the most exceptional intellectual capacity compare to any other *Homo* species on earth. Then again, what is amazing is that these three characteristics converged in one single species. Was it a mere coincidence led stochastically (at random) by natural selection? Was it a unique phenomenon at the universal level which produced rational beings as we are? How natural selection acted to sweep off all other *Homo* species being very similar to each other and similar to us, and we survived those extinctions? Were there any special genetic combinations that made possible our superiority and survival? Was it the onset of a war-aggressive-behavior genetically based in our species that ended the existence of those weaker species?

In our modern history we have been witnesses or readers of such exterminating behaviors and thoughts, displayed by *despotas* as Adolf Hitler and Joseph Stalin. The first, tried to exterminate the Hebrew population conducting the most brutal holocaust ever known in written history, and Stalin massacred whole villages with no mercy, in the name of the Marxist doctrine and the proletariat, that he and a horde of barbarians led for decades. But the worst, they had followers with the same thoughts and behaviors; they shared a common ill idea against people, with no respect to human lives. Medieval times are an example of historical brutality as well: the law of the powerful, smashing the weak to surrender.

One sad example was the campaign the Church led against the peaceful Cathars in Languedoc in the southern part of France. In the first decade of the thirteenth century Pope Innocent III made up his mind to use force against that region, just because in Languedoc Cathars and Catholic neighbors lived as one community. The teachings of Cathars were not welcome by the Pope for they do not adhere to the strict rules and command of the Church, they thought somehow different from the orthodox preaching of the Church, albeit they lived decent and honorable lives, as stated by one knight to the bishop of Tolouse. Pope Innocent called for a Crusade against Languedoc. He promised the crusaders booty and land. According to history, on June 1209 the Crusade was ready for action, with more than fifteen thousand soldiers. The Pope appointed Arnaud Amaury to lead the crusaders. On July 22, 1209, the crusaders began to slaughter anyone in the streets, forced open all doors and massacred elder people, men, women, children, and priests, nobody escaped. Furthermore, crusaders executed Catholics as well just because they lived as neighbors with Cathars as part of the Languedoc community. The Pope and his bishops were perfectly informed of the massacre, but they did not condemn such conduct. Cathars were exterminated just because their thoughts were in some way different from the official Church interests. As in communist countries now a day, at those times Church did not tolerate dissidents; they were labeled as heretics. May all these modern historical facts somehow explain the extinctions of other ancient humans thousands of years ago?

There are several clues in our genetic endowment, such as genes that are "silent genes" producing no proteins for millions of years. Besides, there are genes that have changed spectacularly during the last 200,000 years. Unfortunately, just a small fraction of all possible hominids that ever lived have been found as fossil records, anywhere on earth, which prevent a finely tuned approach to our most distant history, and of course, to essentially touch the sequence of events that marked the big differences among the coexisting species in the past. It is well established by fossil records that our hominid history (I say hominid not *Homo*) is dated back as far as six million years; the first four million years all hominid species lived in Africa, supporting Darwin's hypothesis of our African origin. The ancestor we share with chimpanzees lived around seven million years ago. Besides, the two living species of chimps *Pan troglodytes* (the common chimpanzee) and *Pan paniscus* (the bonobo), shared a common ancestor in the past, probably more than two million years ago.

We may consider the chimpanzee and the bonobo as our brothers, we had a common ancestor; we may not fear the words in evolutionary terms. Some precautious scientists rather say bonobos and chimpanzees are our cousins; no, they are not the children of our ancestor's brother, we shared the same ancestor, the same progenitor species. We share more than 98% of identical DNA.

Almost all belief systems proclaim humility as a requirement to be good persons, we have to accept with human humility that those apes are our brothers, and that those unexplainable turns of "natural selection" acting on different mutations, (that have been detected in researches conducted through the Chimpanzee Genome Project) enclosed them in the jungle, with no hope of constructing a city to live in, as we do.

Another incognita arises. *Homo erectus* became extinct about 30,000-32,000 years ago, possibly later, albeit fossils have not been found for a more recent date. It was a more intelligent species than *Pan paniscus* or *Pan troglodytes*, it was a human; in this particular case development of the brain meant nothing to survival. Chimpanzees and Bonobos are here and *Homo erectus*, an ancient human, is gone. The same happened to *Homo neanderthalensis* (Neanderthals), they became extinct around 30,000 years ago, very close to the extinction of *Homo erectus*. I dare to say, in terms of evolution, both became extinct at a very close historical time. Certainly, this two species were stricken by extinction around the same period. This coincidence might be no accidental, perhaps unknown factors determined their extinction. No many specialists discuss this particular incident in the history of these two species, and if they do, they also avoid going deeper in the issue. Nevertheless, this coincidental disappearance about the same time in history of two well-established *Homo* species is a tantalizing fact to consider, at least from a theoretical point of view, for future studies to establish a deeper insight in the misfortune *finale* of our homologous and better-known species that shared the planet with our ancestors.

The genus *Homo,* which identifies humans, probably emerged between 2 and 3 million years ago, according to most specialists. Records are not available for all those evolutionary steps, and other species of the genus, that probably existed, are completely unknown for us today.

Scientists have proposed not less than three possible hypotheses regarding the expansion of *Homo* species throughout Europe and Asia. Basically, these hypotheses try to explain what fossil records fall short to account for.

The first theoretical proposal was framed during the 1980's and 1990, and considers that very ancient groups of our ancestors left Africa long ago. In that way, without a precise period of migration, *Homo erectus* occupied territories of Europe and Asia. After a long process of evolutionary modifications, populations of this species became *Homo neanderthalensis,* which in turn evolved into *Homo sapiens.*

Geneticists, mainly considering all possible information contained in our own DNA, have proposed the **second hypothesis**. The reader may recall that in a previous chapter an approach was made to the so-called mitochondrial Eve. Once again, mitochondria are the center of interest to elucidate obscure aspects of evolution. The mitochondria bears 37 genes present in its chromosome, but these chromosomes are never deposited into the egg during fecundation, it turns out that for any individual the mitochondrial DNA is nearly a perfect copy of his/her mother's mitochondrial DNA. Every single female will transmit that mitochondrial DNA through all generations, and every male will receive that kind of DNA exclusively from her mother. It is the privilege of females the monopolistic delivery of such inheritance. Hence, her children inherit mitochondrial mutations, and the daughters will transmit that same mitochondrial information to her children and so forth. All these studies based on the inheritance of mitochondrial DNA, point to a true **recent ancestor** in Africa as a second hypothesis.

A **third hypothesis,** which tries to explain *sapienization* around the world, takes for granted that fossil evidence is an irrefutable proof that all the so-called human races are nothing but direct descendants of local ancestors that lived millions of years ago in those specific areas. Beyond that, part of this hypothesis depicts all the people living in the vast region comprising the surroundings of the Black and the Caspian Sea together with all regions around the Caucasus Mountain range, with more than 700 miles in length, as direct descendants of Neanderthals.

Results based on studies accomplished using the Y chromosome, which is exclusively included in the male genetic formula XY, also point to the recent African ancestor hypothesis. In fact, at present, it is accepted that our ancestors originated in Africa in the recent past, (I say accepted but not accurately proved) spreading all over Europe and Asia after the exodus from Africa, going through evolutionary changes, and splitting into different *Homo* species, as has been stated before. As fossils have proved, between 50 and 100 thousand years ago, probably at least three species of humans (genus *Homo*) coexisted on earth, as depicted on Fig. 7.

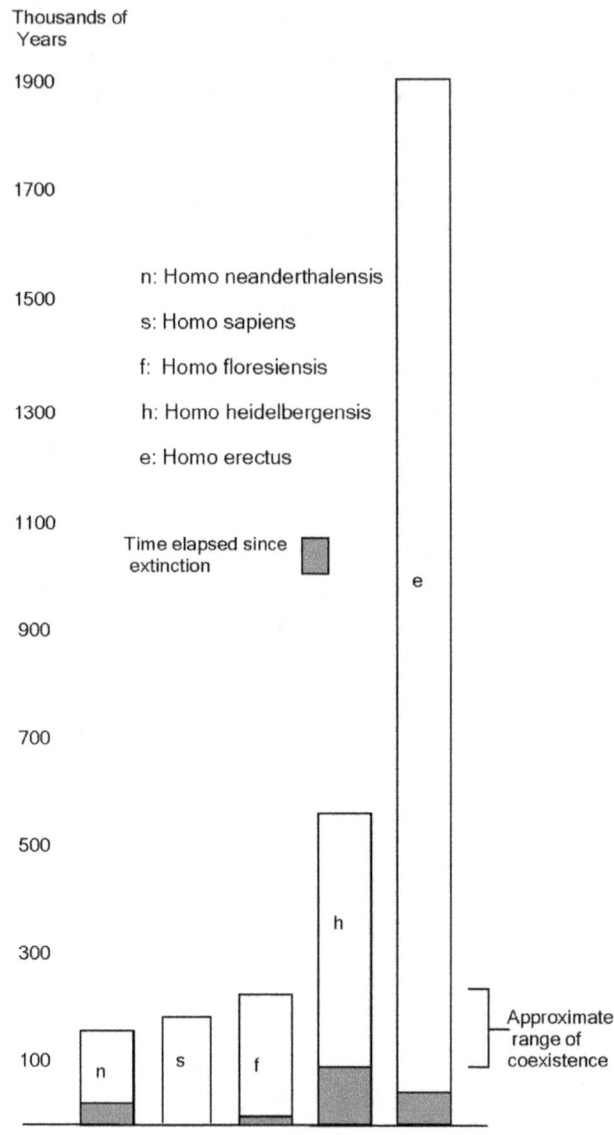

Fig. 7 The five most important human species (Homo) that coexisted for about 100,000 years.

Among them, the most controversial nowadays was an unexpected find in the Indonesian island of Flores, it is *Homo floresiensis*. A very small skull, together with other remains, was found in that said island, pointing to hominids not taller than 3 feet, but completely adult. This new species dated back between 85,000 and 18,000 years old. A very modern species that disappeared as the others did. Compare to modern humans the height of the face is roughly 73%. Some scientists believe they were a pigmy population of *Homo erectus*. They were as modern as *Homo sapiens*. They coexisted with our ancestors, and then disappeared, for no apparent reason; at least a convincing explanation is out of reach by the moment time. This outstanding discovery was announced in 2004. Anatomical features together with brain characteristics point to a close relation to *Homo erectus*, according to other comparative results.

Neanderthals. The Vanishing Kinsfolk

It is legitimate to say that albeit Neanderthals' fossils have been found in different parts of Europe and Asia, it is not obvious how Neanderthals reach their flourishing status; furthermore, neither is clear how the species evolved to thrive the way they did. All we know is that when the "out of Africa" migration led the waves of *Homo sapiens* in their invasive journey to Europe, Neanderthals were there as a sturdy species, muscled conformed, estimated weight 30% over modern humans of the same stature, extraordinarily strong, thriving through an Ice Era that was on the way. Their skulls were low and elongated and their brains as big as *Homo sapiens'*, larger in some individuals, with bulky brow ridges and long face; heavy noses projected forward. Their rib cage, as in apes, sports a bell-shape frame together with prominent broad hips. This broad pelvis probably made delivery an easy physiological process, in contrast to modern human females who not uncommonly face physiological problems at delivery.

It is worth to mention here that between 2005 and 2006, an important number of skeletons were found in Niger, Africa, on the portion of the Sahara desert. These skeletons belonged to two different cultural groups, the Kiffian and the Tenerian that lived in the region separated by 1,500 years in time. Tenerian represent a more modern population, dated between 6,500 to 4,500 years ago. Particularly, the Kiffian skulls showed evident bulky brow ridges, according to the photograph published. Was it an atavism that surfaced in this population as a credit for transitional characters still present in modern humans? Besides, fossils account for stocky people, heavily muscled. They were *Homo sapiens*, since the dating of the bone material placed them from 8,000 to 10,000 years ago. The rest of the skull is compatible with modern people. Based on the above, heavy brow ridges were not an exclusive characteristic of ancient *Homo* species, as Neanderthals and *Homo erectus*, among others. Still, the major number of fossils were found some five years ago and no articles on comparative studies have been published so far. It is an important find comprising anthropology, archeology, and evolution, added to all cultural details that may float up from this material when appropriately studied. Nevertheless, no resonance seems to be born from all this important fossil material.

According to fossils, Neanderthals appeared as a species from 130,000 to more than 200,000 years ago, and then again, it is uncertain the way they evolved and gained the European territory, their homeland. About 70,000 years ago they rambled throughout Europe and into an important portion of western Asia. They settled efficiently on those territories for more than 200,000 years, according to fossil records, which allowed paleo-anthropologists to calculate the time elapsed in their journey throughout those iced lands. The remains dug up by many teams of specialists support the idea that they were a true intelligent species, of course, probably their minds did not work the same as ours since cultural environment, *modus vivendi*, and all the historical background hold by Neanderthals may set them apart from our modern idiosyncrasy. It is a fact, humans as rational beings think as they live, they do not live as they think.

A recent discovery of Neanderthal's tools in Sussex, England, is considered a proof of technological advanced. It appears that these are older than the ones found before in this country and made by ancient *Homo sapiens.* The Sussex discovery is dated 30,000 years old and presumably could belong to one of the last Neanderthals hunting groups, since it is calculated that Neanderthals vanished about that time. According to the specialist that made the report, these tools appear as coming from a population in complete possession of a prosperous technology, not exactly from people at the brink of extinction. To add more, archeologists say that it is apparent the degree of sophistication on these tools.

Other studies clearly state that Neanderthals were as good hunters as early *Homo sapiens,* and completely able to communicate among themselves when *Homo sapiens* arrived in Europe some 45,000 years ago. Furthermore, a team of specialists studied the typical *Homo sapiens'* tools and Neanderthal's dated from the same period, and concluded that Neanderthal's had a slight advantage over *Homo sapiens'.* Their tools were at least as efficient as those from the earliest *Homo sapiens* were. In addition, in 2008 remains of fish, dolphins, and seals in coastal caves in Gibraltar among Neanderthal fireplaces, dating from 45,000 years old as an average, proved the seafood consumption by Neanderthals as well as their ability to hunt for those sea mammals, and their fishing skills.

The entire above are strong evidences against the Neanderthal's extinction based on the tag they were "outwitted" by *Homo sapiens.* Several specialists have expressed it is not consistent with history to assume that the methods of hunting, feeding, fishing and tool use, may account for the extinction of a whole population. Other factors, rather unknown, may explain the vanishing process, the inexplicable swept-off in the case of Neanderthal people, all over Europe and Asia. They were not dumb or stupid humans, evidences have proved they were as smart as early *Homo sapiens* living in the same period, besides being sturdy and well built.

Some scholars affirm Neanderthals possibly evolved from a group of *Homo heidelbergensis* that migrated from Africa to Europe some 500,000 to 600,000 years ago, though this declaration bears a certain dose of speculation; fossils point to such circumstances, and if it was exactly the evolutionary event that created Neanderthals, the process is understood as a gradual and progressive transformation. On the other hand, modern humans evolved from *Homo heidelbergensis* that stayed in Africa. These events clearly affirm that *Homo heidelbergensis* is, in fact, the common ancestor of both *Homo sapiens* and Neanderthals. This could be a fourth hypothesis. Most scientists consider *Homo heidelbergensis* lived from 700,000 to 180,000 years ago. This period corresponds to the Middle Pleistocene. It is almost certain that Neanderthals and our *Homo sapiens* ancestors shared the planet for about 150,000 years. About 45,000 years ago, it is supposed that moderns invaded Neanderthals' hunting grounds and territory.

Neanderthals robustness matched perfectly with the features needed by a hunter. They were strong enough to handle all actions to succeed in hunting. Their sturdy physiques were *ad hoc* adaptations for living in those extremely cold climates, which in turn, are telling us they successfully managed their way of living, to thrive in those inclement weather conditions.

Some few authorities claim that Neanderthals cold-adaptedness is a historical manufactured article, since there is no sufficient evidence of living areas next to the glacial territories. These specialists perhaps are not taking into account that fossil collection is somehow a chance process, hence not precisely evidencing all the possible places where settlements took place. Besides, to deny that Neanderthals physiques are not in accordance with cold region adaptation is to rampantly ignore Allen's Rule, one of the principles of evolutionary biology: legs, ears, arms and any other flange are shorter in mammals living in cold areas compare to close species in the tropics. Eskimos living in arctic regions and Lapps above the Arctic Circle in areas of Norway, Russia, Finland, and Sweden, are vivid examples of the rule.

Neanderthals were successful, no doubt about it. They were our analogous humans, and in some ways better than we are. Heaven knows, if we were direct descendants of Neanderthals perhaps our weak and unfinished DNA would be much better, to prevent so many life-threatening diseases, that we suffer a lot. They stayed on earth more than we have had; Neanderthals dominated the European landscape for 200,000 years, more than four times the time elapsed for true modern humans on earth. They struggled through times that probably could be doubled hazardous for living humans, if we were facing the same conditions.

In fact, to be more precise in saying, Neanderthals' brains, on the average, were larger than ours, to some extent. Being physically strong and intelligent, it is hard to believe that competition or climate could be the influential factors enough to explain their extinction. Pathogens introduced by Cro-Magnon (modern humans around 30,000 years ago) have been evaluated as possible factors for their extinction. Climatic changes have been proposed as well. It is considered that European populations of Neanderthals were adapted to cold climates according to their anatomical structure. As climate changed, moving away from the Ice Age, a selective pressure on Neanderthals was created. A weather trend toward warmer temperatures could cause a mal-adaptation process, according to some specialists. Another possible explanation is Neanderthals' hunting grounds; the forests. As these became basically flat lands during Ice Age, they could not chase their game to stab them as efficiently as they did in the forests. It is worth to say that according to information gathered from fossil records, Neanderthals stab their game;

they did not use spear throwing as Cro-Magnon used to do. Stab hunting required rapid running and their anatomy clearly shows that they had short and sturdier legs than Cro-Magnon, which was an evident disadvantage, since they were urged to get enough nutritious food to sustain their high-energy requirements to survive, if stab hunting was their style. Remember they were 30% heavier than any modern human of the same height. Now, if we think of battles among groups, the Neanderthals's disadvantage is clear on the scenario. These are just hypotheses. Perhaps the true reason is a little portion from them all. Invaders behavior could account for their extinction. At present, fossil records do not show any evidence of serious confrontational encounters between both species that could explain Neanderthals' extinction, as a result. Nevertheless, the possibility exists, albeit is not the only explanation to satisfactorily solve the incognita.

By judging the modern human's behavior, it is not out of the question that arrivals from Africa, probably in several waves as some specialists believe, outcast Neanderthals ended up extinct, just because the outlanders practiced a violent behavior, probably ransacking, killing and devastating Neanderthals' grounds, something very similar to the Spaniards' colonization in Cuba, that sent native *Siboneyes* and *Tainos* aborigines to a rapid extinction, leaving no trace of any descendants, they were completely swept off the island,(as Neanderthals were on our planet) and that happened between 400 to 500 years to present. The brutality showed by Spaniard *conquistadores* was so massive, that very scant elements of the native culture were left to account for the existence of those populations before the atrocious invasion and abuses, of all kinds, perpetrated against the pacific dwellers of pre-colonial Cuba. Something very similar happened in Mexico when Hernan Cortes and his *conquistadores'* hordes took over the Aztec lands, to submissive them all, in the name of the Catholic Queen of Spain. All these and more atrocities happened in modern history, when an allegedly civilization manners would be the mode. It is not an unjustified assumption to think that Neanderthals probably suffered the same treatment from modern

African invaders, when codes of human rights were completely unknown, of course. Europeans, during their roles as conquerors over the Americas, did not practice such codes for centuries either, and all those events pertain to modern history.

It is worth to say that when the first Neanderthal's fossil appeared in 1856, specialists from different countries spilled opinions so contradictory in essence that from a scientific point of view the so valuable fossils came to be something like a "rarity" out of the context of human evolution. As an example of these opinions and contradictions, the conclusions of the German specialist Schwalbe, after a careful study of the remains at hand at the end of the XIX century, proclaimed his scientific opinion was openly opposed to the common belief that Neanderthal was in the lineage of living humans. According to the results he obtained after a careful revision of the said materials, he stated that Neanderthal was a different species, and labeled it as *Homo primigenius*. Others, simply neglected the remains, just because they supposed such fossils were an erratic humanoid form, probably close to Australian aborigines.

The most probable fact that could explain such disparity in the opinions of renowned specialists in science-leading countries was the "ideal missing link," a theoretical form very close to humans with atavist features of pre-sapiens forms. Furthermore, French specialists were in one way or another anti-evolutionist, which precluded any real approach to the evolutionary process of the human race. Besides, Germany and France already had Neanderthal's fossils but England. This fact probably triggered the English specialists disapproval on the evident remains found in those countries.

All these sterile controversies did nothing but deferred a substantial and rational understanding regarding the importance of Neanderthals in the evolutionary scenario of human species. Scientists at that time, as a whole, acted as a braking system against the advancement of paleoanthropology. The one that aimed the biological status of Neanderthals was Schwalbe, mentioned before, as modern genetic studies have proved. He had the clear vision to place Neanderthals out of the living human's lineage.

At present, several results on successfully obtained mitochondrial DNA (mtDNA) from Neanderthal fossils are coincident with the fact that there is a deep split between modern humans and Neanderthals; they stay apart as a different species, which probably happened around 600,000 years ago. Their mtDNA sequence is different from that of modern human's. Their extreme adaptation to cold climates point to the fact that they evolved allopatrically (isolated), precisely in those iced regions where their adaptation became faultless. In contrast, more than 98% of genetic similarity is shared between chimps and living humans, as has been stated before in a previous chapter. Are chimps more closely related to us than Neanderthals? The answer might be the technical procedures researchers used. It is not the same to have results from mtDNA than from nuclear DNA. Studies with chimps and living humans were basically conducted on nuclear DNA. Besides, the fragments of mtDNA obtained from Neanderthal fossils are limited, since not all the sequences were in excellent conditions as to make assertions on a 100% basis. Moreover, as has been stated by some specialists the split of Neanderthals and modern humans was so a recent event, that chances are Neanderthals' nuclear DNA sequences were more intimately linked to any existing human DNA sequences than to other Neanderthal populations. Besides, mtDNA has its own limitations as to be an accurate marker to determine closeness between modern humans and Neanderthals, since mtDNA is only and nothing but a simple portion of DNA. How is, in fact, the rest of the Neanderthals' genome?

Other interpretations have been stated. One possibility is to consider that, effectively, Neanderthals did not provide mtDNA sequences to modern humans, since they became extinct long before any potential admixture. Other plausible explanation is, simply, genetic drift could be responsible for the complete loss of mtDNA in our genome, though Neanderthals were part of our ancestry.

Studies have been carried to conclude that the assimilation model is the most plausible explanation regarding the supposed extinction of Neanderthals. If the model is true, then Neanderthals did not become extinct, their genome diluted into modern African immigrants, giving way to modern Europeans. Neanderthal's genome was assimilated, not replaced. Hence, this assimilation process led to an attenuating course in Neanderthals' contribution to modern humans' gene pool.

Another important Neanderthal fossil material was found very near La Chapelle-aux-Saints, in France, pulled-out by a group of Catholic priests. These important remains were studied and described by Boule, who filled a huge amount of pages to utterly confer the category of "a brutish" to Neanderthals. Many years later it was proved that Boule's descriptions were completely blemished, and misinterpretations of anatomical features were rampant in his manuscript.

All this is just a mere example how science could lead to completely wrong interpretations, just because science has no corporeal or intellectual individuality, science is structured by scientists, and they are humans, with certain intellectual focusing frames, that may or may not be in accordance with the most strict logic based on the facts, and in many cases their thoughts and hypotheses are permeated with individual interpretations on those facts, far beyond the strict reality and context. The worst, some of those well renown specialists stick to their believes and conclusions in such a way that they only expose or state one single point of view, with no alternate hypothesis, which is completely mandatory in good science. They simply say like: "…this is my interpretation according to what I have studied for years." Unfortunately, such attitudes prevent in many instances a true evaluation of the "hard to find materials" that have been carefully preserved by curators for decades, or by geological layers for millions of years. I may call those inflexible criteria as "ghost hypotheses," since they do not reflect the reality behind the biological material, fossils in this case. They keep looking at their intellectual image (the ghost), a creation of their own, as the only plausible explanation of an event that occurred hundreds of thousands, or millions of years ago.

One thing is obvious, if careful studies have been carried on by different specialists and interpretations, at present, are as many as specialists have been around, it comes out that the truth is not so simple to grab on. Sometimes, little pieces of the whole truth might be scattered all over the possible interpretations or hypotheses. But human pride is the exact barrier to prevent a consensus of opinions, and authorities rather stay apart trying to prove that one's hypothesis is correct, being the rest judged as a bunch of disparate statements.

Moreover, as recently as 2009, a careful study has been concluded and published (see bibliography), which discussed and proposed different models and hypotheses to approaching the time of divergence of Neanderthal and modern human lineages. In fact, certain inconsistencies came up when trying to assign Middle Pleistocene specimens to *Homo heidelbergensis* or Neanderthals; the plausible conclusion point to the fact that the scenarios proposed and discussed are not so distant in the expression of evolutionary process. Then, what does lie beyond? Are we still hesitating about Middle Pleistocene specimens when assigning specific names?

It seems paradoxical that humans try to interpret human evolution and fail to accomplish it in so many ways. Many of the former interpretations of Neanderthal's fossils were permeated by the fear of the advancing scientific facts against the official beliefs, in which living humans are an exceptional creation, and nothing "primitive in essence" could be considered a human connection, or related in any admissible way.

They were afraid of the general opinion prevailing in those years; they preferred to send out of the edge the evidence of other humans in the past, than to stand the bitter critics of contemporary colleagues. Besides, scientists in general were engaged with official belief systems, and they avoided any confrontational positions against any religious context not to be pointed as heretics. Beyond that, research methods were not so advanced at those times, and this lack of technology precluded accurate results, so the outcome was a very moderate scientific opinion on something as unusual as Neanderthal's remains, which show themselves as an unprecedented historical fact. Perhaps, this last consideration was of a less importance compare to the rough criticisms anyone could be exposed to, if they dare to go beyond the establishment. In fact, the establishment precluded any statement that could smudge the untouchable truth that "humans are perfect by creation." Such affirmation implicitly carries the arrogant posture that we, humans, are completely apt to understand and speculate on creation, moreover, to evaluate the product of creation itself. This is the utmost show off that we, defective beings, dare to throw on the harsh trail of humanity course.

Finally, other variants of the three main hypotheses already discussed in this chapter, that in one way or another involve the disappearance of Neanderthals, have been proposed as well. One of this suggests that certainly, Africa was the original center of all humans living today. Hence, the modern anatomical features appeared there for the first time, and those genes, were carried by leaving African groups, supposed there were more than one departing group as all evidences indicate. If they found people somehow different, they were not so much different to prevent admixture; which took place. The exchange of genetic material accounts for the intermediate features found in some fossils out of Africa. To support this hypothesis, Neanderthal fossils show a certain protrusion on the occipital bone, which is frequently found mainly in Caucasian Europeans, as Physical Anthropology has demonstrated.

Another variant or hypothetical declaration is that the modern anatomy **did not** go out of Africa, carried by migrating natives "replacing" Neanderthals all over Europe and Asia. Instead, the migratory phenomenon was the gene flow by interbreeding, which conferred modern anatomical features to descendants all over generations. The ending point of this hypothesis is the fact that modern anatomy is skinny compared to the robust Neanderthal's, which sported massive muscular structures and broad and heavy bones.

The advantage, theoretically exposed, of such lean anatomy is fewer requirements in terms of nutrients, since slender bodies need less amount of food, which is an advantage in times of scarcity, and populations as a whole, thrive much better. As a consequence, genes that determined slim lined bodies were positively kept through the action of natural selection, and all the way through many generations they changed the *façade* of the *Homo* body, becoming the advent of modern human's anatomy.

It is not necessary to bestow any kind of superiority to migrating Africans, since the advantage of modern anatomy rests on the fact of saving energy from less food consumption. On the other hand, the evident problem that arises with this theoretical variant is that a faint proportion of Neanderthal genes have been found in modern humans, according to some genetic studies. Then, the genetic endowment of Neanderthals was completely replaced, albeit only anatomy related genes were supposed to be selected, together with a pool of genetically based biochemical processes aimed to accomplish a coordinated physique-biochemical course, granting a harmonious biological result, as the modern human is. But the result was a complete replacement of Neanderthal gene pool, as said before. As stated in a previous paragraph, evidences abound nowadays to credit the assimilation model, not the replacement one.

A plausible explanation is that there are many genes inherited together as a "gene complex," which could explain the complete replacement to some extent. Nonetheless, the precise reason for the total replacement of Neanderthals' gene pool is not clear whatsoever. It is by no means possible to completely understand how so many Neanderthal populations disappeared all of a sudden, since the evidence provided by fossil records demonstrate that their vestiges were gone around 28,000 years ago as the latest, which is only explainable if some cataclysmic phenomenon really occurred, which is not the case, since no geological or environmental evidences are available. Was that cataclysmic event, if any, only catastrophic for Neanderthals? Modern *Homo sapiens* kept on their evolutionary path rather rapidly, which is a major contradiction in terms of extinction regarding two very similar species. We have to disregard this biased supposition in the name of good reasoning.

Other speculative propositions based on scarce and simple details tried to picture Neanderthals as not organized hunters but furtive ones, and as opportunistic scavengers, all of which is an evident detrimental approach to intelligent populations of *Homo neanderthalensis*. The hearth and remains found in Gibraltar, discussed somewhere before in this book, account for a very different approach. Moreover, fossils from Mount Carmel, Israel, obtained by digging in four different caves provided another scope on Neanderthal populations: Kebara, where important and typical Neanderthal remains have been found; Tabun, remains of a Neanderthal female with no so extreme features; Qafzeh and Skhul, where classic modern characteristics are present in all fossils collected so far. Being all these caves in the same area, they grant strong evidences of coexisting Neanderthals and modern humans, together with stone made tools with no different manufactured structures, implies not to set those populations apart on a technological basis.

These clues situate the issue as no differences, since neither group provide evidences of new mental capacities. It was a fact that what fossil records indicate is that Neanderthals and moderns went "side-by-side" but never interbred. It is impossible at present trying to understand how was the impression of a Neanderthal female in the eyes of a modern human, or the other way around. Perhaps they were really reluctant to mate, and interbreeding was just a rare and sporadic event. The above discussion comprises the Levant area, as the reader may notice. This geographical region has been considered of a great importance regarding the evidence of practically neighboring groups, of modern humans (Cro-Magnon) and Neanderthals.

The non-interbreeding issue though both groups were evidently nearby in a small area, comprising less than 8,700 square miles, could be explained by certain isolating mechanisms that fossil records cannot account for. The Neanderthal physiology is totally unknown since fossils do not hand out any clue, not even to guess a little about such characteristics. Modern human females are always ready to mate no matter what the estrus cycle might be. We cannot be so sure about Neanderthal females either displaying the same behavior regarding their reproductive cycle, or any other difference on the subject of mating behavior itself.

If there is a strong isolating mechanism among species, in a general sense, it is the reproductive isolation, which may be expressed in many ways, physically and physiologically, with slight details or important ones. As we are completely ignorant on the reproductive characteristics of Neanderthals, we have no other choice but to speculate on an issue that almost certainly will never be clarified. The general consensus on the issue is: they did not interbreed. This is another blur point in Neanderthals' history, which account for the perception that they were a different human species. If Cro-Magnons at that time had a self-conscious as beings, recognizing only their congeners as equals, is another point that could be taken into account as a possible explanation for the reproductive isolation; but this as well is just a possibility since we are not in good knowledge to enter the fascinating psychological world of Cro-Magnons, our closest modern ancestors.

Not only it is difficult to believe two coexisting human groups in a real small area with sexual isolation, it is also astounding that there was no competition either during 50,000 years living as neighbors, at least that is what some specialists believe. Then again, fossil records do not offer clear clues on the competition subject; in any case, no proofs of violent encounters are entangled in the remains of those caves that could justify a harsh competition due to quarrelsome groups, pushing each other outside.

If we accept all this, as fossil records show and dating techniques reveal, then a very simple conclusion arrives, they tolerated each other. Then again, why Neanderthals disappeared? Particularly, in Mount Carmel, this question brings the mind of every scientist that had been working there to a closed road. Are there any particular or special attributes in that area to fully explain such an incredible event? All data available from the area squeeze out to deny the well-known and pragmatically proved principles of species competition and niche resources. Is the Levant, particularly Mount Carmel, the corner stone to really understand why we are here? And if it is, why was it? What was the fact or facts that disabled the successful course of Neanderthal's evolution in every single area all around their distribution? Natural Selection does not seem to be the answer. At present, there are no evidences that could point to any particular trait as a biological target for Natural Selection to act upon and shot down Neanderthals. Most important, Neanderthals were scattered almost over all regions of Europe, Israel, Northern Arabia, Caucasian area, north of Black Sea, and far beyond the eastern coast of the Caspian Sea, at Teshik-Tash, as the extreme eastward spreading point, according to fossil records, and probably, other areas where fossils have not been found. Within all these different climatic and geographical conditions, what was the common selective factor that sent them to extinction? What was the nature of that factor, or factors that in spite of different climatic and geographical settings smashed Neanderthals populations to the brink of extinction, and finally consummated that task?

Precisely, based on those remains from the Levante it is very clear that Neanderthals and modern humans behaved in a very similar way, not distinguishable one from the other around 65,000 years ago. If we take this as a true historical fact, then it becomes that Neanderthals and *Homo sapiens*, at that time, had identical brain functions. It is, they were the same intellectual beings into different physical models. Along the way, in an uncertain time in history, presumably a drastic neurological change occurred that altered *Homo sapiens* behavioral pattern, then, a new blueprint of behavior was installed in turn. How this change took place? Speculations, smart guessing, propose hypotheses or simply inferences should be used to approach the unknown. From this time on, we may consider the second split regarding *Homo neanderthalensis* and *Homo sapiens*. At that very moment, the split process determined two different intellectual trails for both species.

The said split probably occurred between 670,000 and 120,000 years to present, as has been proposed by several scholars.

Time-Border for Neanderthals

The Upper Paleolithic period, according to some specialists, might be explained by a "biological event." It is theorized that a sudden neurological event occurred 50,000 years ago that led to completely new cultural advances. According to this approach, brain changes were the booster to achieve all developments that followed. If only one archaic human lineage acquired these advanced neurological features, chances were that they spread all over. On the other hand, no increase in brain volume, no evidences of any other kind of changes pointing to differential anatomical brain structures, according to fossil records, appeared whatsoever. If this hypothesis would have any chance to be fitted to updated finds, the advancements in artwork, tool technology and alike, must point to specific geographic locations; but this is not the case, at least, fossils do not evidence this. What the scenario shows are scatter-like events, which comprise modern fossils, as in eastern Europe, thousands of years before than in any other location. Art showed up here and there, with gaps of thousand years as well. Extreme examples of *avant-garde* technology or any other culture demonstrations have been found scattered with the same pattern. Neanderthals burials proved to be carefully crafted, in disperse locations. Furthermore, beads and small hole-drilled teeth have been found in Neanderthal's remains.

Some authors claim that the out-burst of culture behaved in an epidemic-like fashion, spreading almost all over Europe in just few thousand years. Of course, different styles of artwork have been unearthed from one place to another, as an evidence of local creativity and development. How did this explosion of cultural progress take place in short elapsing historical millenniums? A group of experts ponder the acquisition of language as the undeniable trigger to explain such a spreading phenomenon in a relatively short evolutionary process. Against this hypothesis there are **undeniable** proofs that language probably appeared a million years before the cultural explosion took place.

According to meticulous studies performed at Columbia University, it was found that inner skull prints demonstrate *Homo habilis* possessed the capability of language. This accomplishment gives the perception that any *Homo* species cannot be considered speechless, in a more or a lesser degree. Those prints show the left-brain hemisphere development in perfect agreement with Broca-Wernicke cerebral areas, which provide the capacity of language. Language was there long before the art and technology appeared in a bursting fashion. Humans were endowed with speech capabilities, which seem to be the reason we may call all of us, *Homo*. Of course, to conclude beyond any reasonable doubt that *Homo habilis* mastered language needs to be proved and fossils are not enough to sustain such an affirmation regarding a primitive species as *Homo habilis* was. At Max Plank Institute for Evolutionary Anthropology, preliminary results in the reconstruction of Neanderthals' genome, shows the possibility that they had the same allele of gene FOXP2 that modern

Homo sapiens have; this gene is connected to speech capacity. The existence of our *genus* preceded any art or technological developmental process; our ancestors were well bequeathed to continue the unstoppable transit toward what we are today. Our species out stood itself from the very day "evolution process" accomplished the final touch, to put our ancestors apart from the rest of the living world in our planet; even apart from our brothers, the chimps. *Homo sapiens, primus inter pares!*

It seems to be absurd to match tool technology progress with language alone. Many tribes during the first part of the twentieth century in Africa and Australia were nothing but hunter-gatherers, besides sporting a fluent language with profuse communication. Furthermore, nowadays there is a group of hunter-gatherers well known in Tanzania, located along the southern part of Lake Eyasi, in the Great Rift Valley, they are the Hadza. They are not involved in any kind of agriculture, not even the most basic one. They do not have a permanent shelter, besides, no livestock at all.

According to genetic studies, they could be an example of the main ancestry of the human family tree, probably more than 10,000 years to present. Nevertheless, as they do not live densely enough to risk an infectious epidemic, or to suffer famines, it is not exaggerated to attest their lives run in a fair healthy condition. Probably, natural selection has done a very meticulous job to provide a well-adapted gene pool for Hadza people. Their diet is more fairly balanced than the average diet of city dwellers all over the world. Such results prove that famine episodes, or meager food supplies menaced hunter-gatherers neither in modern times nor in remote periods. An important cultural aspect to be mentioned is the fact that they do not show a deep grief when one of their own dies. There are no funeral services or rituals of any kind; no grave markers at all. They simply bury the body and go away; no matter how close the dead person was.

What some authorities overlook is the fact that as far as 35,000 years ago human populations were but small groups scattered over huge territories, with no many possibilities of encounters. The most probable event was not a meeting to exchange technological experiences but confrontational stumble upon each other, since residents fear to be invaded, and then lose their hunting grounds. Under these evident conditions, how could language may spread technical experiences or artwork accomplishments? That is out of the question. The capacity of language worked so much better as an intra-population possibility to pass experiences, no matter what kind of, from one member to another, and from one generation to the next. Just take a look at Medieval History, and you will find such behaviors as intra-group communication ability and distrustful attitude toward new comers, as the rule. Moreover, imagine a group of marauders around your neighborhood. What your concern might be? We are modern and civilized humans after all, aren't we?

On the extinction of Neanderthals, writers and authorities have proposed thoughtful approaches to the issue. According to fossil scatter-patterns in several locations, evidences point to the fact that Neanderthals were not adept walkers, since their stocky bodies with rather short limbs did not enable them to be ramblers in the same way as long-legged Cro-Magnons were. They presumably only moved from one place to another in search for food to sustain living. Under these conditions, every evidence point to the fact that males did not provide children and females with food, or at least not enough for their requirements. Females search for food as roots, edible leaves, very small game, insects and any other kind of safe to eat meals. It was completely possible for females, since they also sported sturdy physique, to struggle for existence and provide their children with food and care. In fact, Neanderthals seemed to lack a family structure as we modern conceive the subject.

Results from studies conducted on the need for nutrients for Neanderthals to strive, point to the fact that based on their sturdy body frames and muscle mass, they needed a meat diet, plenty enough daily. To accomplish this, not only males were engaged in pursuing large games, females did it as well. Accordingly, Neanderthals roughly consumed 54% more kilocalories per day, to be well fed, than an average modern human. This is what results from studies conducted on dietary characteristics of Neanderthals, showed. The outcome of all this was evident. Information obtained from studies accomplished by comparing skeletons of juvenile Neanderthals and moderns, living during the same period, clearly point to contrasting results. Almost 50% of Neanderthal adolescents died, not arriving to adulthood; on the other hand, fossils of modern humans on the same period, account for only 25% of deaths, or less. Besides, though it is not completely clear how the silent estrus became an attribute of modern females, chances are that Neanderthal females were receptive only during perceptible indication of estrus. Modern females of the Upper Paleolithic were ready for sex anytime, giving more strength to the pair bond since their hidden estrus made them apt to accept sex not conditioned by a reproductive cycle at all. That stronger pair bond possibly became a protection pattern for the family group, something that lack the Neanderthals' family structure, or so believe.

If the low rate of juvenile survival was a true pattern during twenty generations or so, together with the fact that family structure did not protect females from harsh struggle, and reproductive cycles were ruled by an unknown estrus frequency, then Neanderthal's populations stalled and then vanished due to impoverishment of healthy conditions in their populations, reproduction, and the survival dilemma. In those Neanderthal declining generations due to the aforementioned happenings, the logic event was that modern Upper Paleolithic humans, eventually outnumbered Neanderthals.

At this point it is worth to say that interbreeding could be a fact, since the decreasing number of Neanderthal's populations and the small number of their members, did not provide a contribution in terms of genes to be an evident proof of such admixture; Neanderthal genes were "diluted" through more than one thousand generations ever since. Of course, these are only assumptions that authorities state to give some plausible reasons in front of Neanderthals' extinction. In fact, we can take these proposals as acceptable hypotheses that need to be confirmed.

The main problem remains as a certain impossibility to test each one of these hypotheses; fossils do not account for everything needed to arrive to a flawless theory of Neanderthals disappearance. Why all these drawbacks became a real menace 30,000 years ago and not before if Neanderthals managed to co-exist with moderns for more than 10,000 years? What was exactly the event that drove Neanderthals to fade-out from the human stage? What kind of selection and on what traits made possible such a relatively sudden vanishing? It was an all around event, since not only certain localities witnessed the farewell to that human species. The phenomenon occurred throughout the distributional Neanderthal area, as has been stated before in previous paragraphs. To my best understanding a possible explanation is the high mortality rate of juvenile Neanderthals, as a very serious study already proved, which precluded the normal growth of populations, though this fact alone is not strong enough to exterminate a whole species all around its distributional area, which was in fact a large territory with different ecological characteristics. Specialists have to look for other explanations to convincingly determine a more plausible approach to make clear Neanderthals' extinction.

Some authors have stated that Neanderthals were unable to compete with *Homo sapiens,* and those authors try to explain Neanderthals' extinction based on this arguable disadvantage. If they really were in a disadvantageous competition, it does not mean that it was the ultimate reason for their extinction, since native South American are still alive though high technologies and huge cities are all around, besides Spaniard conquerors settled down in those lands centuries ago; the natives keep their culture almost intact.

Neanderthals were not facing invaders that held a superb high technology in warfare terms, which could be the worst of all calamities they could stand. Besides, the world at the time was not as crowded as it is now. There were enough places for everybody, and perhaps this is one of the most important ecological aspects that most authorities overlook. Neanderthals were not under siege, there was enough "elbow room" to any population, our planet was under-populated, and there is no question about it. That was a fact; no one has to guess on this. Furthermore, just 10,000 years ago nearly 9 million people lived on our planet, from North to South Pole. Then again, at the time Neanderthals disappeared the number of inhabitants on our planet represented a very small population.

The harshness of climatic worsening conditions made food collection by females a cruel task, where no sufficient nutritional meals were accessible for everybody within the group, at least, in cold regions where most of Neanderthal groups were inhabitants. Decreasing the number of reproductively active females and adding the fact that, males available to mate were few since old individuals were part of the group, led to a border-line-limit of reproductive success, a bottle-neck, which finally pushed down the birth-rate, together with a low survival rate of juveniles already happening. Both events occurring in most populations gave way to the species decline, perhaps with other factors acting upon to exacerbate the already critical species survival. Extinction was imminent.

I need to add something else to all previous discussions. If we, living humans, do not share any important proportion of genetic material (genes understood) with Neanderthals, this means that our genome and theirs are different. Then, a very simple conclusion arises to indicate that our genome is not an

exclusive endowment to be humans. Neanderthals had another genetic formula and they were humans as we are, besides being very successful. Furthermore, if nowadays DNA analysis could confidently be applied to *Homo erectus, Homo heidelbergensis,* and *Homo ergaster,* the most probable outcome would be different genetic formulas for those species, too. Furthermore, as has been stated before somewhere in this book, living humans and chimps share more than 98% of genetic similarity based on both genomes. They are apes and we are humans. Neanderthals were humans not apes, and there are no inherited genes from their genome, at least, as an important contribution. This is somehow astonishing; an ape is closer to us than another human species. Chimps are genetically so close to us that we share blood groups too. This contradiction is so overwhelming that, on one hand, more than 98% of genetic similarity is not enough for chimps to be humans. On the other hand, we do not share Neanderthal's genes, at least as pointed by results from mitochondrial-DNA studies; different genomes account for both *Homo sapiens* and *Homo neanderthalensis* as two different human species. Unfortunately, to have Neanderthals' full nuclear DNA available to perform any kind of comparative study seemed unlikely. According to certain results, Neanderthals and modern humans split after some 460,000 years before present. The last remains of Neanderthals are dated around 30,000 years ago. The possibility of useful nuclear DNA was just a dreamed possibility; we had to rely exclusively on mt-DNA, which is only a small fraction of the genome. Recently, on May 2010 at the Max Plank Institute for Evolutionary Anthropology in Leipzig, Germany, new sequencing technologies made possible the analysis of some million DNA portions of Neanderthals' nuclear genome. Then, it came into light that Neanderthals and *Homo sapiens* interbred to the point that up to a 4% of their genome is in ours. Previous research in 2008 was only based on mtDNA, which is very limited information. Of course, this 4% is not so important as an inherited proportion, but it definitely proves that both species were not so distant in social and cultural way of living. Besides, this genetic proportion also proves that viable individuals were born from those out breeding; we could not possess that 4% of Neanderthals' genes otherwise.

If we look at this from a true neutral position, we have to conclude that the creation...evolution...emergence of human beings is not based on one single and unique genetic model, many human genetic variants, probably many more than fossil records account for, were on earth living, and feeling, and struggling for life, and suffering in many different ways long before our *genetic model* was ready to be what we are today. There are many genetic models for being humans. This incommensurable scenario looks like a colossal experiment on rational life, where the lab is planet Earth. Steps to create humans were a graduated route of a well natural oriented purpose, where every new species overcome the previous one by adding or modifying all shortcomings that its *passé* demonstrated as a flaw. Was it natural selection all the time disabling *Homo habilis, Homo ergaster, Homo erectus, Homo floresiensis,* and *Homo neanderthalensis* just because none of those previous species models did any good? Was always there a natural reason to exterminate those human populations and, hopefully, make possible the emergence of another one, theoretically, in a better shape to succeed in spite of the same adverse conditions or even worse? It may seem that natural selection is an utmost lab technician manipulating every single detail to help to "create" an intelligent, social, industrious, active, lovable, creative, and as irascible as an ape, human beings that we are.

What is next? Will there be any other *Homo* species to replace *Homo sapiens* in the future, perhaps not far from now in evolutionary terms? We do not know any other *Homo* species on earth; this is not a possibility. *Homo* species' natural history has been the replacement of one by another, or a trivial coexistence period with scarce contacts, if any. Would it be different for us? Is our species already evolving to a better socio-biological status, so fast and drastically in terms of an evolutionary period that a new genome will take over, and the current, outcast, as a major event? Do we have enough time for that? What are the genetically based traits that natural selection will favor next, for us to turn into a new and better species? Will the next step be the DNA configuration that we have been longing for? We can be sure we are not meant forever. None species has ever been. We cannot crave for that.

Unfortunately, equally amazing events occurred regarding another *Homo* species that settled down in Asia and Africa. Next, I will approach that human species whose fossils account for a long lasting permanence on earth, more than any other *Homo* species ever known. Likewise, this species vanished with unclear facts; why that happened that way? It is another divisive subject in the convoluted history of humans' evolutionary process, another genetic formula to set a human being on the stage for natural selection to act upon, and then swept it off.

Homo erectus. The Long-lived Human Species

In 1891, the young Dutch anatomist Eugene Dubois dug out a skull cap in the eastern part of Java. Next year he found a femur together with other skeletal remains. He published a paper (1894) on *Pithecanthropus erectus,* giving the idea that his find was an ape-man, so intriguing the missing link of those days. A hot controversy burst-out since some anthropologists considered *Pithecanthropus* nothing but an ape. Others incline their opinions toward ancient human remains. During the 1930's von Koenigswald decided to explore, on a methodical basis, the so believed fossil beds in Java. Fortunately, he obtained many additional specimens, not only skull caps. Finally, specialists recognized *Pithecanthropus* was not the appropriate genus, since after a careful study performed on the fossils, it was evident to place the species in the human category, hence genus *Homo* was the appropriate one, becoming *Homo erectus* ever since.

During the 1960's, it was widely believed that *Homo sapiens* is a direct descendant of *Homo erectus*. Nowadays this misconception has been abolished, being the current tendency to consider *Homo sapiens* a direct descendant of *Homo heidelbergensis*, as has been discussed in this chapter before. *Homo erectus* fossils from East Asia were dated from around 2 million years ago to almost 30,000. The expansion of this human species is believed to happen between 1.8 to 2.0 million years, probably from the southern part of the African horn, starting at some point around Tanzania and Kenya, then moving northward to Egypt and splitting in three different directions:

a) Through Egypt bordering the eastern coast of the Mediterranean Sea through nowadays Turkey, Bulgaria, Northern Serbia, Austria, and then northern Italy.

b) Through Egypt, northern Arabia through the land between the Black and the Caspian Sea.

c) Through Egypt to Iraq, Iran, Afghanistan; around this land they split again into two different ways, the northern procession through People's Republic of China, and the Southern parade through northern India down to Thailand, Malaysia and Indonesia.

These geographic descriptions are just approximate according to fossil records and logical inference from all the gathered evidences up to date. There are different criteria regarding the taxonomic status of *Homo erectus*, since some experts consider that many of the fossils judge as *Homo erectus*, in fact, belong to different hominid species. Others claim that *Homo erectus* was a single species that occupied territories in Africa, Asia and Europe. Due to its large distributional area, local differences are found in fossils from different geographical regions, as is completely usual in widespread species with populations located in dissimilar zones.

Albeit most specialists do not coincide with the fact that standardized Neanderthal and *Homo erectus* skulls are quite similar, though Neanderthals' are bulkier, the general profile, on both, is perhaps closer than skull profile in living human's individual differences. I myself have met a person sporting a well-projected occipital bone very close to any Neanderthal known today. He was a normal person, but that characteristic was notorious in the neighborhood. He was a child around thirteen with a very short haircut, which made visible such exceptional skull outline.

It is my impression that some specialists forget the normal variability found in any species. Individuals of any animal species are not clones, they are genetically unique, hence show variations that anthropologists sometimes forget about, perhaps just because to stumble upon a fossil is an easier said than done event, then the notion of variability is not important *per se*. Evaluate the fossil in its concrete features is the utmost goal, no other considerations may be pertinent because with fossils it is rather difficult to get into population genetic practices, which is the *ad hoc* Genetic discipline to establish a variability status for any species under revision. When a biologist arrive to the ample conceptual criterion of species variability, then science is in perfect shape to adjudicate the morphological, physiological, behavioral, and ecological limits within which that specific species ought to be considered, we are risking to be splitters or lumpers otherwise, and it is one of the most detrimental defects in Systematics for it dangerously sneaks out from the truth, leaving behind the impartiality that science deserves.

It is possible that a revision of certain fossils might be mandatory, always with open minds, taking into account that life is variability; genetic variation is a law, not a simple tool that one can take or drop according to individual conveniences or preferences. Any biologist, even the most modest student has to be formed in the principle of genetic variability, which is the truest visual phenomenon of life itself. There is neither life without

Octavio Perez-Beato Ph.D.

variability, nor clever conclusions on fossil materials disregarding variation. The correct systematic status on so important fossils may lead to a better and easier understanding of human evolutionary history and variability. Misconceptions may lead to a disastrous disappointment, by facing the impossibility of a harmonized analysis that permits a close approach to the real evolutionary events that brought-in and discarded so many different human species in the past.

On December 1929, a first skull was found at the Zhoukoudian cave, some 26 miles from Beijing. Two years before, in 1927, only one tooth was found and based on this single piece a new genus and species were named, *Synanthropus pekinensis*. Such a designation tended to comply with the so long expected "missing link" or ape-man. A popularized name soon was applied and has been well known all over the world as the "Peking Man" ever since. It had not been yet identified as *Homo erectus*. The digging up of the skull reassured the taxonomic status of *Synanthropus pekinensis*. The scientific reconnaissance was not at the time a true understanding they were facing a hominid species, closely related to living humans. In 1934, the expert that published the record of *Synanthropus* died at work.

Several skulls, different bone structures, and fragmented jaws and teeth, were found at Zhoukoudian, China. Unfortunately, the original fossils of *Synanthropus* were lost forever while the turmoil of World War II was shaking China during the occupation by Japanese troops. The real fate of those precious fossils had never been known. Fortunately, casts, photographs, measurements, and detailed written records have been kept, all of which made possible all the scientific literature being published to let the world know about the finds at Zhoukoudian. Besides, scientists involved in this study soon became aware of the extreme similarities between the Peking Man and the Java Man, whose fossil were found in the Island of Java some 50 years before, and then named as *Pythecanthropus erectus*, as the reader may recall. The relation was evident, but at the same time it was

- 132 -

astonishing to conceptualize a *Homo sapiens'* past relative located in two so distant places. At the time, the evolutionary process of humans was not so clear and many misconceptions prevailed, nothing odd, since today many aspects of the humans' evolutionary process have not been clarified either, at least in satisfactory terms.

Years after, both Java Man and Peking Man became *Homo erectus*, then recognized as the species that left Africa, an amazing event not well understood at present, and only daring surmise may be used to approach the most extraordinary episode in the primeval human history. Of such a dimension that said departure was as to open the possibilities to invade a significant part of the world by the creature that, as unique as it was, plunged into the vastness of new lands to become the superb *Homo* species that left strong evidences of its peerless longevity, more than 1.8 million years on earth, not equaled by any other hominid known today, of course, we included.

Homo erectus most striking characteristic is the thickness of the skull bones, as has been discussed by many specialists since the first basic descriptions published many decades ago. The skull cap, in particular, looks like a massive low dome or tortoise carapace, as has been described by others. Open discussions and evolutionary interpretations have been debated to explain the extreme thickness of the skull. Almost all arguments point to a defensive-aggressive purpose, set up by natural selection, probably also connected with sex-competition, according to some authorities. No other *Homo* species bear such a helmet, not even the modern apes. On the average, the parietal skull bones in *Homo erectus* are almost three times as much as in *Homo sapiens*; in fact, it is an armored protective structure to securely encase the brain, not found in any other primate.

Based on the ample existence span of *Homo erectus* as species (covering almost 2 million years), natural selection could accurately act favoring such skull measurement, only if socio-behavioral experiences, in *Homo erectus* populations, gave way to situations where dangerous threatening brain-damage was involved. An immediate suggestion appears according to the above avowal, it is, an aggressive recurrent behavior could be the regular group demonstration, in many instances, during the lifetime of any *Homo erectus* individual, perhaps daily.

In 1984, in a Kenyan region along the Nariokotome River a whole skeleton assigned to *Homo erectus* was dug-up. It was the anthropological working team of the well known scientist Richard Leaky the fortunate group that brought into light such an outstanding discovery. This skeleton is known worldwide as the Nariokotome boy, since it was determined, after a very accurate study that those bones belonged to a child about 10 years old. Calculations demonstrated that the boy was about 5 feet and 3 or 4 inches tall, a considerable dimension for such a young individual. The estimated full-grown tallness was calculated above 6 feet. All previous considerations based on the scarce sample of bones obtained from *Homo erectus* remains, predicted a moderately size species. This is another proof that scientists, sometimes, speculate on aspects that posterior facts prove wrong. Nariokotome boy proved to be as tall as a tall modern human, when full grown.

Now, in possession, for the first time, of a whole *Homo erectus'* skeleton, which provided invaluable information to all specialists involved in human evolutionary study, some authorities believed that it was more proper to assigned the skeleton to a *Homo erectus'* ancestral species, called *Homo ergaster.* Nevertheless, most researchers keep the former designation. Again, criteria may diverse among specialists since a complete and definite understanding of the species is impossible, since bones cannot tell about patterns of behavior, or group social skills, or even degree of language development. Bone variations may mislead opinions, in one way or another.

Studies on every single detail of the skeleton continued, and surprisingly the vertebral canal was narrower than expected, in the mid dorsal spine segment. A conclusion burst out almost immediately…"Homo erectus did not master language; the species could not talk." Why did they suppose that *Homo erectus* had to have the same level of language development as we modern humans have? We are talking about a dated material of about 1,500,000 years old. It is impossible to think of modern speech in those ancient populations. One thing is the similarity in physical body and another very different is to master speech as modern do.

Only if a trait is genetically based and its manifestation is obviously present in a sufficiently repeated manner creating at least a phenotypical dichotomy, in many cases more than a variant of the character is involved, natural selection can act upon selecting the most convenient one according to the existing environmental conditions. In this particular case, the socio-behavioral environment where *Homo erectus'* life happened to be. As an example, the evolution of the other important hominids, including *Homo sapiens,* lost forever the thick skull trait, discussed before. Then, does it mean they were less aggressive than *Homo erectus?* Was the change connected to the style of "aggressive-defensive weapons" an influential factor to transform the thickness of the skull bones in the rest of *Homo* species? In what measure were Neanderthals, *Homo heidelbergensis, Homo floresiensis,* and *Homo sapiens* less aggressive than *Homo erectus?*

We are aggressive in general terms, much warfare in human history account for that issue, in modern history and in ancient history as well. Sophisticated massive destruction weapons, to smash the opponents in modern times, have been used and we do not sport a thick skull any longer. Then, after all these considerations one idea, in particular, comes to be obviously logical and acceptable: most of the confrontational attacks between and among *Homo erectus* individuals were directed to bash the head of the opponent, in so a frequent fashion that natural selection could act upon selecting the thickest skull, for them to survive, in those perilous times of savage hordes. Probably, that was the way *Homo erectus* came to have such an amazing thick skull bones. Still, the main question remains unanswered, why did not we inherit a thicker skull to better protect our precious brain? Is it because the increasing brain volume needed a thinner bone shell that could evolve rapidly as a suitable brain case? How was it possible that the increasing volume of the brain perfectly matched, in time, with the increasing size of skull bones to encase the cerebral mass in a harmonious and perfect content-container manner? If skull bones did not get larger at the same pace as cerebral mass did, an obvious conflict, in evolutionary terms, would arise to dangerously compromise the development of such an important organ as brain is. It has been reported that skull caps in very early hominid fossils show a thickness comparable to that of *Homo erectus*. Then, what a perfect coordinated assemblage to produce the enormous leap that the enlargement of the brain means in human evolutionary history! Natural selection may act with all the power of selective forces based on any genetic trait, but it is too much to expect coordination in organ assemblage and development based only on natural selection; if we do, then we are taking for granted that natural selection can do anything and then, it shall turn to be a reductionism concept, since all possible explanation will always converge in just one single and simple phenomenon, natural selection.

Thanks to genetic studies two genes have been discovered that are connected to the formation of the human brain. Different alleles (gene variants) are known to be carried in different frequencies (percent of people bearing the allele) depending on the world region. It is believed that other genes may be involved in the development of the human brain as well, comprising different alleles with specific functions on the size of that organ.

Sometimes, and this discussion is the case, the outcome of evolution is astonishing and so incredible that, what we call **natural selection-evolution concatenation process,** seems to be an inexplicable course of events, hard to believe as a spontaneous progression occurrence, at all. The emergence of those genes connected with brain development has to be explained by obvious mutations. The point is that there is no plausible way to explain why those mutations appeared, just in time and in the correct species to advance the process of humanization. That/those mutation(s) or any other similar(s) did not occur in other primates.

Certain facts seem to be clear about the limitations and possibilities of *Homo erectus*. It is believed that at least a limited or basic control of fire really existed for *Homo erectus*, though it is not very well understood the limits of performance on this so important skill, evidence appears that it was. No tools have been found to assess that they were hunters, every evidence points to the fact they were scavengers that probably used fire with certain limitations to cook meat when possible. Stone tools were very primitive, probably used to scrap meat from carcasses as left over from carnivores' meals. This does not mean they were not gatherers; the most plausible idea is, they were scavenger-gatherers people. In fact, this detail places them in a primitive hominid condition. In addition, strong evidence indicates they were not well adapted to cold climates, rather to warmer lands, which point toward a moving-season-nomadic style.

Of course, fossils are not so abundant as to be a real demonstration of the behavior associated with nomadic season-oriented people, though it was probably the most likely moving pattern for their groups. Nevertheless, *Homo erectus* was an unfailing species that actually was successful in its time, regardless of certain limitations imposed by the evolutionary history they had. As it was said before, it has been the *Homo* species that endured an existence-span of nearly two million years, under severe climatic conditions among other harsh circumstances. So successful were they that at a certain period in our human history, *Homo erectus* was one of the species that coexisted, for a short period, with early modern humans. It is hard to imagine if such coexistence would linger beyond the limits it was in the past, what ethical principles would be appropriate to apply to sibling species (*Homo sapiens* and *Homo erectus*) in our modern world? If they were not as modernly evolved as *Homo sapiens* were, chances are that slavery or oppression would be the result of the evolutionary disadvantage. Slavery regretfully happened in our history within the same species, with no differences in the evolutionary path. But not only slavery shamefully occurred, in medieval time kings and queens subjugated people that were not descendants from the aristocracy.

Language is still a controversial issue among different specialists, including neurologists. Anatomical aspects have been considered to determine the real possibility of *Homo erectus* mastering language skills. Different hypotheses have been proposed, but all those theoretical approaches are deficient in one important aspect or another; hence, not a true and consistent consensus has been reached at all. Nevertheless, something seems to be obvious, *Homo erectus* did not produce an articulated language as modern humans'; probably they communicated with much more efficient sounds than living apes do. Some authorities speculate that probably their language was somehow below the level of Neanderthal's language structure, and of course might be considered a very primitive language, with major guttural sounds with poor or no inflexion at all. If they were not hunters, then language with a certain degree of development was not a necessary tool. The reader must recall the arguments regarding the narrow vertebral cavity on the mid-dorsal segment, discussed in a previous paragraph.

Since coordinated movements during the hunting activity are necessary, language has to be present. There is a strong interrelation hunting-language as a major complex to obtain food. It is very difficult to conceive even the most basic hunting strategy without the use of language. If communication, during those critical moments of hunting down a prey, is ineffective, the activity itself turns to be extremely dangerous or otherwise completely inefficient, mainly if we consider big mammals. Perhaps this behavioral aspect, hunting, could explain the lack of developed language to some extent, if *Homo erectus* was not a hunter species at all. Carnivores can run at high speed to grab down their preys but humans cannot.

Presumably, the rest of the group activities could run without the need for a developed language as modern human possess. Gatherer and scavenger activities are far simpler than hunter's is. If in the future any archeological find proves to be a tool or a spear-like instrument intended for hunting associated with a *Homo erectus* site, then we have to seriously reconsider all previous assertions, which denied the virtue of language for *Homo erectus*. Of course, we cannot and should not conclude that only grunts and yells were the sounds that their primitive language was tuned to construct. In fact, it is very difficult to ascertain the extension of their "diapason" and the complexity performed in the emission of sounds. On the other hand, if they really fairly produced fire, that very act implies a coordinated reasoning focused toward a concrete objective.

At this moment, language specialists know a lot about the rational development in languages, comprising a fairly acceptable range of millennia, used by our species (*Homo sapiens*) but nothing could be determined based on fossils. In any event, the group cohesion through a steady understanding and cooperation could be crucial for the survival of the species, which linger more than a million years through different evolutionary stages.

Fossils prove and authorities agree to consider early and evolved types of *Homo erectus*, in Africa and Eurasia. Such a fruitful species, through a considerable time span in the evolutionary context of hominids, had to master, at least, the most basic performance of a primitive language at that time, developed enough to allow the proper communication to guarantee cohesion and cooperation, *sine qua non Homo erectus* would never had been the successful species it was.

Another stigma commonly used to characterize *Homo erectus* is the well known, and discussed by many specialists, evidence of cannibalism based on different fossils. Nevertheless, there are two possible explanations for the practice of cannibalism: a ritual performed based on certain cultural beliefs, and the imperative need to alleviate the famine, which might not be an unusual group reality in those harsh times they lived. Besides, cannibalism is a well documented event in certain tribes dwelling deep in the jungles in different parts of the world. Not only cannibalism, but the custom of shrinking the heads of enemies has been a practice of a certain tribe in the Amazonian forests. Fortunately, that practice was abandoned some seventy years ago.

Primitive behaviors are sometimes difficult to understand at the light of our civilization, as it was astonishing for Hernan Cortes' troops when they observed human sacrifices in Aztec's altars devoted to divinities. All these events were very well documented in the book written by Bernal Diaz del Castillo, one of Cortes' soldiers.

Homo erectus relents. The pathway to extinction

Recently, in 2002, a new hypothesis (model) was published trying to explain the possibility of populations' replacement as has been discussed elsewhere in this book. This hypothesis postulates that the extinction of a population is followed by the onset of a neighbor border population (recolonization process), and if this phenomenon is repeated over hundreds of thousand years, evolutionary trends are imposed and genetic changes do occur, then the overall perceivable effect is a total and complete replacement of one species by the other, when in fact, only multiple recolonizations were achieved. The recolonizing populations bear almost the same genes as the previous replaced one, but not exactly identical and genetic variation increases as the recolonization process is repeated through time.

The reader may recall the "Out of Africa" replacement model, in which a new *Homo* species totally replaced Neanderthals and *Homo erectus* as well, since at that time the coexistence of *Homo neanderthalensis* and *Homo erectus* has been well documented by fossil records. Somehow similar, but a different hypothesis has been proposed, termed as the **clinal replacement model**, which conceives a low and steady population transition based on a gradual fashion, not as abruptly as the replacement model ascertains.

The *Homo erectus'* population in Java, it is believed, was isolated for a longer period, and probably with almost no gene flow for almost one million years, since the intermittent low and high sea level prevented, during a long period, the possibility of gene flow from the main land *Homo erectus* populations; hence, a somehow different phenotype for Java population was almost certainly evolved, according to some specialists. Some two million years ago Southeast Asia, Sumatra, Borneo, and Java were all connected as a broad mass of land known as Sunda, which made possible the supposed migratory route of *Homo erectus* to Java. At that time sea level was as low as 80 meters or lower, than it is today.

It is hard to calculate, if not impossible, how large as a whole, was Java *Homo erectus'* population since fossil records are useless to make such calculations; in my opinion, any attempt to estimate population size, is very speculative. Precisely, based on the impossibility of any population size estimation, it turns out that it is completely hypothetical to propose that modern human invaders outnumbered *Homo erectus* in Java, as the reason for its extinction in that particular island. In fact, as several authorities have acknowledged, chances are that Java populations lingered near to 48,000 years ago, which made possible the previous mentioned coexistence of more than one *Homo* species during a certain time in the past, albeit their potential contacts remain unknown. Different authors have mentioned the possibility that *Homo erectus'* populations cohesion may indicate that they cultivated a tight group behavior, which rendered protection for all the members, and developed the practice of sharing, which, in the end became the keystone of the first human settlements. In that way, the socially expanded behavior in modern humans was planted long before the emergence of the most ancient forms of *Homo sapiens*, which in fact, entered the modern human evolutionary stage around 195,000 years ago, as fossils indicate.

Fossils in mainland China point to the fact that *Homo erectus* was replaced rather slowly by early *Homo sapiens*. Besides, some experts express that in Java *Homo erectus* was replaced by early *Homo sapiens* as well, though the arrival of our species to Java was delayed in hundreds of thousands of years compare to mainland China. Nevertheless, Java ecological and physiographic characteristics preserved a relic population of *Homo erectus* (barely 40,000 years to present, or less) up to the time *Homo sapiens* met them, for their annihilation. The main problem arises when accepting the hypothesis of *Homo sapiens* in China evolving from *Homo erectus*, a kind of *in situ* evolution embraced by Chinese anthropologists; in contrast, authorities stated that in Europe and Africa *Homo sapiens* evolved from *Homo heidelbergensis*, which in turn evolved from *Homo erectus*.

All of the above create a genetic gap in the evolution process regarding *Homo sapiens* in China. As a consequence, *Homo sapiens* in China never ever inherited any biological traits from *Homo heidelbergensis*, which was the direct ancestor of *Homo sapiens* in Africa and of Neanderthals in Europe, as is commonly accepted. It is, *Homo sapiens* evolved from *Homo heidelbergensis* that stayed in Africa, and *Homo neanderthalensis* evolved from *Homo heidelbergensis* that left Africa and migrated to Europe. According to some authorities, a corroborated time-overlap involving latest *Homo sapiens* and earliest *Homo heidelbergensis* in mainland China has not been demonstrated. Nonetheless, the fact that fossils have not been found to make an obvious line of descendant *Homo erectus-Homo heidelbergensis* in China, does not mean that the event never occurred. Finds are circumstantial events, depending on geological-climatic conditions that make possible fossilization processes.

Here again, we are facing deep discrepancies in the onset of *Homo sapiens* in Asia, since it is by no means completely clear the, sometimes, uncanny connection with *Homo erectus,* since fossil records, genetic analysis, and anatomical contrast seem to be incoherent to some extent, to make obvious the true scenario where both species were playing the main role. The presence of *Homo heidelbergensis* in China is confirmed by fossils, however, the role of this species in Asia is not apparent as had been in Europe and Africa. What exactly happened? How was it possible that modern humans evolved from *Homo erectus* in China, if for the rest of the world, as the most acceptable hypothesis postulates, *Homo heidelbergensis* was the undisputable common ancestor of *Homo sapiens* and Neanderthals? The assertion of such *in situ* evolution of *Homo sapiens* in China dumps the theoretical possibility of a far east *Homo sapiens* and a western *Homo sapiens,* which is completely improbable. Then again, the lack of proper fossils to account for the intervention of *Homo heidelbergensis* as the immediate ancestor of *Homo sapiens* in China, create the mirage of *Homo erectus-Homo sapiens* direct line of descend. To erase *Homo erectus'* thick skull bones from our modern genome, it was needed a deep transformation in terms of evolutionary process. It is not only the fairy tale of the conversion of one species to another highly advanced, and anatomically different regarding skull and face bones as well as other anatomical structures, it is so much more.

As a consequence, when a group of respectable authorities do not arrive to a consistent consensus, it appears that the truth, the valid point of the matter, has not been reached at all. *Homo erectus* as well as *Homo neanderthalensis* are evident incognitas in the thorny trail of human evolution It is not a matter of being agnostic or not, science try to do the best possible, circumstances delay or misguide the perceptions, and factual evidences usually are hard to grasp. At this point, it is worth to say that certain characteristics in Neanderthals' skull are very similar to *Homo erectus*. Both had elongated cranium, occipital protuberance (occipital torus), and heavy brow ridges. Of course, not exactly identical, nonetheless it is not hard to deduce the common ancestor relation for both species. On the other hand, modern humans split from Neanderthals at a certain time from the ancestral form *Homo heidelbergensis*, which expansion occurred about 650,000 years ago. The exodus started from some point in Ethiopia through Sudan to Egypt and then Turkey, at this point they split in two groups, one moved bordering the northern coast of the Black Sea to Poland, Germany, and France, and ended at some point in the Iberian Peninsula. The other group went across Kazakhstan, probably bordering the northern coast of the Caspian Sea, to the eastern coast of China, most likely near the Yangtze River.

One of the most important fossil sites was located at the northern part of Spain, in the Atapuerca region. Mainly two sites are located in that particular region, yielding fossils since some 30 years to present. First, The Pit of Bones, it is a rather impressive cavern pothole starting at the most profound alcove of the Atapuerca Cavern itself. It goes deep dropping to approximately 50 feet. There, remains dated from 300 to 310 thousand years have been found. The other site is Gran Dolina, where fragments of a child skull were dug up. These fragments are quite similar to the Nariokotome boy's skull structure. A new name was given to these remains as pertaining to a new species, *Homo antecessor*. In fact, according to other specialists this "new species" seems to be nothing but another *Homo erectus* population with local characteristics of adaptation, as may occur for any species with a wide distributional area, as it happens nowadays in human populations in any country around the world. In some descriptions the Atapuerca population consisted of people with heavy brow ridges, stocky and muscular physiques, and rather tall. Some scientists believe that Atapuercans could be an evolutionary branch toward Neanderthals, since finally a whole skull could be recovered and studied in full details. All evidences point to the fact that those transitional human forms really settled down at Atapuerca around 800,000 years ago. In fact, Atapuerca region seems to be a glimpse on the evolutionary process from *Homo heidelbergensis* to Neanderthals. Fossils in this region represent perhaps the "instant transition," in geological terms, to tell us that Atapuercans were neither pure *heidelbergensis* nor genuine Neanderthals, but a biotype between both, hence one of the most important halfway evolutionary path in our human history.

Modern humans, *Homo sapiens,* did not and do not bear those skull characteristics as Neanderthals did. Our skull changed completely when we emerged from *Homo heidelbergensis,* if we really did. Why did those structural genetic traits endure in Neanderthals and disappeared from us? Is it an evolutionary law when two splitting species emerge, always one of them is less anatomically advanced than the other one? The same happened when an early hominid split from chimpanzees. They are still apes, and we are highly developed human beings...the winner takes it all!

The Leap into Humanness

Ancestral human lineage parted from apes some six million years ago, then, through an unknown event difficult to explain and only possible to infer, those early humans somehow developed behavioral patterns that set them apart. One of those, perhaps the most important, was moving away from the forests, while chimps stayed there. How did those early hominids look like when time came to split from chimpanzees? It is hard to answer. For most specialists they were nothing different from australopithecine's form, or much closed to chimps, though the fact of upright walking is still on the arguments. Yet, they bore the genetic information to evolve in a different manner, away from the chimps and away from any other primate at the time. Their genetic variation, as a bequest in their genomes, gifted them to start an unprecedented course of evolution, hazardous, faltering, facing all imaginable hard experiences to finally emerge in the anatomical shape, intellectual development, and social complexities that delineate what, we, human beings are today.

Perhaps, about a million years went by long before they became something like *Australopithecus afarensis* or *Australopithecus africanus*, our forerunners, some 3.5 to 4 million years ago. But even so, hominid brains grew no larger than chimpanzees', not before 1.8 to 2 million years ago, according to some authorities. The complete and unique differentiation process took a long journey. Then, what environmental pressures acted upon those early hominids for them to quit the forests, presumably, while chimps never did? A possible explanation rests on the fact that some five million years ago there was a severe drought in Africa; as a consequence, a rather substantial part of the forests turned into woodland patches, being a possibility that such an environmental change, irremissibly, pushed those early evolving hominids to the flatlands. On the contrary, chimpanzees found the way to stay in the scarce forests, and thrived there.

Was it, in fact, the importance of an environmental pressure or a very special genetic endowment, as a primeval spark, for early hominids to become completely different from apes? If by any chance, those *ad hoc* genes to support living in flat lands would not exist at the moment of the environmental changes onset, what the future of the human species would be? Does anybody dare to answer? Then again, the proper environmental pressure on the special genetic endowment simply produced a biological outcome, with unpredictable significance: humans, as we call ourselves.

Many authorities have grabbed the hypothesis of "out of the forest" as the first step to become bipedal creatures, which in turn led to the hominid-human condition. In contrast, the baboons are primates that left the deep forest some millions years ago, and they never ever gave up being four-footed primates. At least, in the case of baboons the selected "out of the forest" kind of living and to stay on the ground decision, did not lead to a bipedal condition at all. What were the circumstances on those unknown ancient hominids that apparently represented the primordial generation in the long evolutionary chain toward humans? Furthermore, roughly, there are four ways to move around in the case of primates. First, most monkeys move on the tree tops, using brachial movements to almost fly from one branch to another, or quadruped on branches as old monkey species do. Second, apes mainly use knuckle-walking when on ground, and branching on tree tops as any other common primate. Third, members of the *Australopithecus* genus (*australopithecines*, as a general hominid group) they walked upright on ground, as any subsequent hominids, but they sported long arms, brawny, to freely move on tree tops as well, or simply, long arms could be an ancestral remnant almost useless for australopithecines. Fourth, from *Homo sapiens* down to archaic hominids, all movements have been exclusively bipedal, occasionally climbing trees as any modern human kids may do.

We may say that these four moving styles characterize what the evolution has done for primates to move around in their environments. The case of bipedalism is just a particular evolved mode of traveling, achieved only through a complex evolution of different anatomical structures that, in an incredible manner, became perfectly synchronized since *australopithecines* era or long before. The point is not only or exactly the phenomenon of bipedalism itself, the point is the extraordinary coordination in the evolution of all the anatomical involved parts, from the brain to the joints, to generate a walking creature. Natural selection decides what trait may go on but, what mechanism is in charge of the perfect coordinated assemblage, including brain capacity and performance of so many structures, at a time, to produce such an outcome?

Dating back some 1.8 million years as the oldest fossils prove, another important human form known as *Homo ergaster* existed, as an additional difficult step in the path of the evolutionary process toward a more complete human model, as we are supposed to be; according to fossils, it probably would have been as tall as six feet or more. Their legs were long, their arms short as ours, teeth also small and the skull drastically changed with a small lower jaw as we have. With small toes and its ribs forming a barrel-like contour, not the funnel shape observed in apes; they were very close to what modern humans look like. Of course, the proportions and features of their face were not exactly as the modern *Homo sapiens*, and the volume of its brain was about 850 cubic centimeters or a little bit larger compare to *Homo sapiens'* 1,400. The trend is to consider *Homo ergaster* as a predecessor of *Homo erectus*, but factually this is just inferred according to some fossil evidences, but is not completely corroborated. Its arms, though brawny, were human's length, which indicates that trees were not an important part of its environment; they seemed to be mainly identified with the African flatlands.

Some specialists consider the reduction in male *Homo ergaster* size as an indication of some important social changes, because apes show a considerable disproportion in stature between male and female, due to a separate hierarchy for each sex. Though *Homo ergaster* females were smaller than males, the difference is not as ostensible as in apes; hence, this may indicate that some new aspects in *Homo ergaster's* community were present. It is believed, and generally accepted, that this was the first *Homo* species with naked skin, without the furry coat that apes have. It appears that *Homo ergaster* is in the direct evolutionary lineage of *Homo sapiens*, according to the most accepted evolutionary tree for *Homo* species. Besides, it seems that, for the first time in evolutionary history, *Homo ergaster* possessed a projected nose, which undoubtedly conferred a more human-like appearance to its head. If both, *Homo erectus* and *Homo heidelbergensis* split from *Homo ergaster* (Fig. 8), and it was a naked-skin species, it is not erroneous to consider the possibility of *Homo heidelbergensis* as a naked-skin species as well, adding to the fact that this species is considered *Homo sapiens'* ancestor species. Again, our species was not the first to sport naked-skin as a privilege that set us apart from the furry apes, other humans, long before *Homo sapiens* came on the stage, showed that modern characteristic. According to certain studies, it appears that the loss of hair (keratin protein gene) in humans occurred some 240,000 years ago.

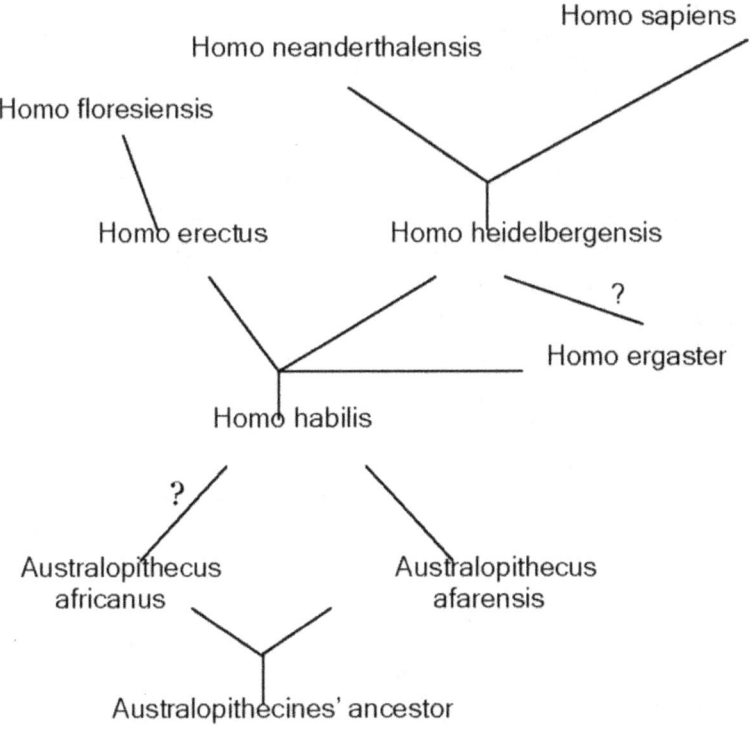

Fig. 8 Proposed evolutionary tree for the genus Homo

Once again, the human model that we are is by no means exclusive, feature after feature has been added to complete the "modern human model" that, one hundred years ago, was believed to be unique, peerless and above all creatures on earth. We are not different from other species; our exclusive privilege is the fact that we are bipedal-walking, full-speaking, and rational animals…to some extent.

Homo ergaster was adapted to live in different environments, based on this it is believed that this species is one of the first to leave Africa, as has been demonstrated by fossils, reaching both poles of the continent about one million years ago. Yes, we may say that this hominid species was relatively closed, in the overall form, to the modern human look. This species is not so well known as Neanderthals. Nevertheless, they were humans and disappeared, with no convincing explanation for such an event. They seem to evolve from *Homo habilis* around 1.8 million years ago, with unknown transitional forms that were likely to exist; no fossils available. The same is true for *Homo erectus* (see Fig. 8), another blown human species that lasted beyond any other mark. Only *Homo heidelbergensis,* known from Europe about 550,000 to 400,000 years ago, went straight as the predecessor of *Homo sapiens* lineage, according to most authorities. In 1907 a jawbone was discovered in the town of Mauer, Germany, and years later a tibia and two incisive teeth, from the same species, were dug up in Boxgrove, England. Said mandible was remarkably thick and the teeth were rather small compare to such a massive bone.

The tibia length and its structure indicated that the individual was an adult probably over 6 feet tall, fairly muscled, comparable to Neanderthals. Later, other specimens were known from Africa, Greece, and France. The skull bones were surprisingly thick as they were found in *Homo erectus*; still, cranium was larger than those of this latter species. In fact, *Homo heidelbergensis* has been described based on very scarce material. Darwin was very clear stating: *Hence, in determining whether a form should be ranked as a species or a variety, the opinion of naturalists having sound judgment and wide experience seems the only guide to follow.* (The Origin of Species; Chapter II, Variation under Nature).

Fossils do not seem to indicate, unequivocally, that *Homo ergaster* evolved to become *Homo heidelbergensis* from which split both Neanderthals and *Homo sapiens,* as has been stated somewhere before. Nevertheless, some experts have affirmed, as a possibility, that *Homo heidelbergensis* is a direct descendant of *Homo ergaster.* The main problem anthropologists are facing is the lack of fossils to attest the complete validity of *Homo heidelbergensis* on the evolutionary stage of the genus *Homo.* It has been said by different specialists that if a complete skeleton could be found, as was the case with Nariokotome boy, the huge gap in understanding *Homo heidelbergensis* could be solved once and for all. Unfortunately, only smart guessing is possible, hence we have to be very cautious when considering this species as the root from which Neanderthals and modern humans evolved.

The "most accepted" evolutionary process for the genus *Homo* is schematically depicted in Fig. 8. In this picture, the time elapsed in the evolution of the species has not been evaluated; the only purpose of this graphic information is to show, in a simplified manner, the evolutionary relationship of the genus *Homo* as most currently accepted, or supposed it happened. Nothing is conclusive since the gaps that inevitably exist, prevent any irrefutable statement on the complicated, and not well understood, evolutionary occurrence of the genus *Homo,* comprising an unpredictable amount of species that had not been recorded as fossils.

In fact, the amount of unknown *Homo* species may be a vital clue for the overall comprehension of the evolutionary trail of humans. Furthermore, a real possibility is the disappearance of many transitional forms simply because geological conditions of the places where those forms died were not appropriate to preserve the remains of most of *Homo* species. Then again, science does not possess the complete picture, and probably will never do. It is not a mere agnostic position or declaration, facts are eloquent enough to easily forecast what the progress of knowledge on this issue shall be in the future, and more gaps than evidences is the rule. It does not mean that future and important discoveries would not be possible; on the contrary, chances are that some peerless results will be on hand in whichever places experts decide to explore. The point is, it is my opinion, that we may expect two possible outcomes: very important fossils are yet to be unearthed, or the best has been already discovered, and only scatter and not highly significant remains is what smart expectations point to.

Chances are that among those unknown species lies the clue to open up the undisclosed legacy; then, the real evolution is forever hidden. Furthermore, to definitely draw significant conclusions pointing to the genuine evolutionary process, only just a minor number of complete fossils, are available. Whole skeletons, or at best, almost complete ones are the exception, not the rule. The rest of the scientific process is composed of a series of inductive reasoning, from the particular to the general concepts, to consolidate a body of principles that may explain the course of events. Not even at the molecular level, as is the case with mtDNA, answers would be better, since any fossil older than 100,000 years, is useless to obtain appropriate DNA samples.

Then again, this is not the first time this postulate is stated in the text; when so many experts from very different disciplines are trapped in restless arguments trying to shed some light on a certain matter, the most probable situation is a serious lack of information as fossils are scarce, or the subject itself is of a tremendous complexity, probably beyond the possibility of present knowledge, or a combination of both, which is more complicated on its own nature.

At present, it is not very clear the relation between *Australopithecus africanus* and *Australopithecus afarensis*, the progenitor species of *Homo habilis*. It is worth to say that both above-mentioned *Australopithecus* species disappeared from the fossil records, and then *Homo habilis* appeared with no other hominid intermediate forms. Forcibly, we have to believe that certain forms between *Homo habilis* and Australopithecines existed, but fossil records failed to preserve such important hominid species. Furthermore, in Ethiopia's Afar region, where Lucy was found, a sloped terrain known as Hadar shows the astonishing fact that *Australopithecus afarensis* is found and suddenly disappear from the geological records, then a lapse of 200,000 years appears as empty of fossil records, then again, suddenly *Homo* remains show up for the first time after this long period. What really happened during the long lasting lapse? In fact, the true and complete evolutionary tree of the genus *Homo* is mostly provisional, since the unequivocal relations cannot be guaranteed, though the technological advances to determine the age of fossils are really impressive, and the competency of very accurate DNA technique analysis are remarkable. Discussions are opened among many specialists, and when reading different books and scientific articles, one finds that every single author introduces certain variations when explaining the evolutionary sequence of events. Recently, two skeletons of a new species, *Australopithecus sediba*, have been unearthed in South Africa, dated less than two million years old. This new species is supposed was roaming near to present Johannesburg, at that time. Both skeletons show features of australopithecines as well as of those of *Homo* species. When studies on this new find have been completed, new aspects of our own evolution could be at hand. According to scholars involved in the study of these new fossils, *Australopithecus sediba* is a descendant of *Australopithecus africanus*. All evidences point to the fact that the evolution of humans followed a very slow process, comprising an evolving pace made of small steps with a series of hominid forms, from which this new species is an important clue. With this discovery grew the evidence that a considerable unknown hominid species is the most probable scenario to reasonably explain human evolution. Humanness was not accomplished with just few hominid species in a row. Many intermediate forms and sibling species almost certainly conformed the family tree for the genus *Homo*. What seems most plausible is the fact that australopithecine forms played a significant role in our thorny evolutionary process. They incarnated the biological path *sine qua non* our species, *Homo sapiens*, had never been a reality. What is completely unclear is the transition from australopithecine forms to the first *Homo* species. Only fossils, that would be available in future expeditions, could shed some light on this intriguing process of humans on earth. If these fossils could not be found, the true and complete course of the extraordinary evolutionary process of human species will never be utterly known.

The brain volume in *Homo habilis* was about 700 cubic centimeters on the average, and this species appeared some 2.6 million years ago coincident with rudimentary stone tools, but those tools did not follow any precise design, simply they prepared and shaped the stones the best way they could, following a very basic practical random design. According to archeologists, this style was prevalent with no ostensible change for almost 850,000 years of history. In fact, albeit this species is included in the genus *Homo*, its main features are very close to australopithecines, sporting an ape-like form in its general appearance and far from the advance *Homo neanderthalensis* and *Homo heidelbergensis*. For many scholars, *Homo habilis* represents a crucial step from advanced australopithecine forms towards the most rudimentary forms of the genus *Homo*.

Two characteristics pushed the scale of the taxon *Homo* to be applied to this hominid species: its brain volume together with small teeth. Both are considered *avant garde* attributes, which place this species in a very privilege path toward humanness. As depicted in Fig. 8, facts and evidences indicate that *Homo habilis* was not the potential predecessor of *Homo ergaster, Homo erectus,* and *Homo heidelbergensis*, the latter being in the direct line of ancestors to *Homo sapiens*.

As some specialists consider *Homo ergaster* just as a local variation of *Homo erectus*, hence not well granted as a different species, and *Homo heidelbergensis* with scarce fossil remains, it is difficult to ascertain the placement of species in the *Homo* lineage. Nevertheless, in spite of all these disagreement, if the previous consideration is entirely correct, then it turns out to be that *Homo habilis*, in its ape-like humble profile, could be important in the emergence of our species...our emergence. Nonetheless, some authors consider *Homo heidelbergensis* and *Homo erectus* evolved from *Homo ergaster* just about 1.7 to 2 million years ago.

Now, it might be obvious the anthropological importance in collecting more fossil materials of *Homo habilis*, since its placement in the evolutionary tree is not clear enough due to the scarce material available. It is believed that this species survived in African lands up to 1.7 million years ago, as an average. Some new features are evident; its brain was developed to 680 cubic centimeters. In contrast, they kept long arms and short legs, which may indicate, as some scholars suggest, that they were still tied to tree climbing. In spite of some primitive characteristics, stone tools have been found associated together with remains of this species, which is an evidence of its skills in manufacturing tools. This quality put them apart from apes though on a body-shape perception they were somehow close to *Australopithecus* species. What if *Homo habilis* would never be on earth? Would it be possible for *Homo sapiens* to emerge? At present, there is not an appropriate answer from a paleoanthropological point of view, since the position of *Homo habilis* is somehow undetermined. Which *Homo* species would possibly continue the evolution toward higher forms without *Homo habilis*? Was *Homo habilis* a dead end species? If all these events would occur in a different way taking into account all the variants mentioned before, probably the face of our planet would be very different as we see it today. Every single step led to the outcome of our species. Was it a stochastic process that could happen in any other way? Think about it, as a possible variant in the evolutionary context, among many others that we could not observe in the moment of occurrence.

The status of *Homo habilis* has been the target of much debate too, since its ape-like profile and the assertion of long arms, very much like australopithecines, has immersed anthropologists in an uncomfortable position accepting *habilis* as specific for the genus *Homo*. Things have been so extremely discussed that there are specialists considering *Australopithecus habilis*, not *Homo habilis*. Others claim for more bones, and mainly skeletons, to really determine the evolutionary status of *habilis*. Things might be very simple, and in fact only a very subtle transitional degree of evolutionary progress is all that set *Homo habilis* apart from the genuine australopithecines. Of course, the closer is one to each other the more difficult to determine the status of different species based only on scarce bones, scattered over different ecological areas. Then again, variation might be another possible issue, since *habilis* could be as variable as any other species, why do we have to expect a rigid morphological pattern in every single unearthed bone? They were not clones; they were simple species with all the attributes of secular variations as any other common species, as field biologists, all over the world, know that happens in all animal populations because of sampling work, besides genetic drift. If we add all these to the previous **subtle transitional degree**, then the picture of *Homo habilis* and australopithecines become a smooth linkage phenomenon, common in systematic zoology. Why do some paleoanthropologists, together with other specialists of related disciplines, try to focus on human evolution as a different biological issue from the rest of the zoological scenario?

Based on available data, authorities may deduce, with a certain degree of certainty, the theoretical relations amongst the different *Homo* species. Besides, in Fig. 8 not all *Homo* species that are acknowledged, based on fossil records, have been included, only the most significant in the evolution of modern humans. Moreover, as stated in previous pages, many *Homo* variants probably existed in the past, not fossilized, hence not susceptible to be known from fossil records. Besides, the branching phenomenon pertaining to our evolutionary tree would never be completely known, as not all hominid species that played some role in that long transcendental journey to humanness, since they remain as fossils to be unearthed.

Along the difficult trail of human evolution some few important things appear quite clear. There have been many "natural projects" of humans on earth that scientists have included in the genus *Homo*; it is, in the family of humans, family *hominidae* as a classificatory group in the zoological taxonomy. They have no other choice; we have to be classified as any other species has been so far. Besides, the fact that other humans, as genuine as we are, have become extinct is undeniable. The reasons for those extinctions, unmerciful extinctions, is supposed to be basically due to environmental changes that the species could not stand, they were not genetically endowed to respond adequately to such changes. In contrast, climate change has been call to mind as a fundamental fact in human evolution. Actually, science is a long way from a complete identification of the binomial combination, climate change-human evolution.

Natural selection favored *Homo sapiens* as the peerless model of a human product, though we are not perfect at all. We lack a lot of characteristics that a finished human model deserved, but natural selection failed to select them properly, or perhaps, we never carried such genetic possibilities since we are imperfect from the very beginning. Natural selection never had the chance to do it better. Then the question is, why such an imperfect model could survive, and strive, and spread all over the world with such deficiencies? This experiment is difficult to understand from an evolutionary point of view.

We are not a privileged species based on our genetic endowment. Then again, we are not and will not be the superb species on planet Earth. We have a lot of advantages, it is out of the question, but in a fair balance we have as many flaws as advantages. We are very far from being perfection. We human beings are plagued with countless inherited problems in our behavior, not to mention our health status. We are still in a hard route of evolution. But due to modern times that route is harm-fully impacted by our own civilization achievements, which in many cases have a boomerang effect on the advancement toward better ways for humans and their fate. We have reached many extraordinary successes in the progress of technology, from the simplest electronic devices to the most advanced space ships, also considering the architectural colossus our countries sport. But together with and not independent from all these achieve-ments, we are slowly walking toward an irreconcilable position of human dwellers of high developed cities and environments, with the manifest intolerance toward other human beings, disregard-ing the essence of humans. Arrogance, selfishness, ill-power, dis-dain toward others, indifference, criminality, and finally social isolation on an individual basis are the worst nowadays evident attitudes, increasingly obvious in hundreds of societies in our modern world.

We live together in highly populated cities, but we do not share enough to be a community of advanced human beings. It is just like certain archaic behavioral patterns are still present, in a considerable proportion, in our human population. It appears that certain behavioral ancient patterns still remain in our gene pool, in open contradiction with our modern way of living. Prob-ably, we still have ahead a long journey of evolutionary pathway to approach a real civilization status. High technology does not exactly mean high advanced social levels in human populations. Not when a certain proportion of that high technology ought to be used to stop criminality, or to subdue other people. Today we use weapons much more terrible and destructive than the swords, spears, and muskets used in the XVII century. We have heavily increased our power of destruction together with the commodities of our high technological world. We have reached tremendous levels of technological sophistication, but they are not well balanced with our achievements in genuine behaviors of high biological beings. We are too young a species to possess pristine behavioral patterns to confirm ourselves as highly civi-lized creatures. We are not yet, we are far from being that model.

Moreover, technology itself could be of detrimental results when used without all precautions that demand certain technological tasks. A vivid example is the regrettable spill of crude oil that occurred in 2010 in the Gulf of Mexico. It was devastating for the entire ecosystem, reaching destructive dimensions for thousands of maritime species, including birds that make their living in those tropical waters. Unfortunately, there are thousands of examples to prove that we are the most destructive species on Earth. We are not the owners of this planet; we are just dwellers as any other species is. We must not make decisions using technology that in the long run could affect entire ecosystems. Those damages may become irreparable. We are destroying our home, which is globally understood, our wonderful planet. Our own technological advancements and achievements are becoming a boomerang. We are not handling our progress in an adequate manner. We are risking the future; we are risking our survivorship as a species.

Our forerunners were deprived of any technology, they were very primitive, and as helpless as a newborn, taking into account the difficult day-by-day life they faced, struggling for existence in all the ways it had to be done, if survival was the goal; it was the motto they carried on, with no other choice. In fact, they were a newborn species compare to our intellectual capacity, social, and biological development. No matter how many thousands of years they rambled on earth, their evolution was very slow, so slow that biological and intellectual requirements were not on time when environmental circumstances push them, as species, on the way to cruel subsistence. None of them were really prepared, on a genetic basis, to succeed in so hard ecological tests. Actually, natural selection was unable to follow a straight way; on the contrary, based on fossil records it appears that a tortuous path was the only handy procedure to produce a "well finished" human being, but we are not at all.

It seems, at the light of present knowledge, that the manufacture of humans was in fact one of the most difficult tasks natural selection faced on our planet. Is it because we reign on animal kingdom? The lineage to modern horses was not so difficult, according to official science, based on the sequence of fossils obtained. It appears just like evolution was discontented with the previous *Homo* models, and decided not to quit until the target model was really achieved. Suddenly, out of the blues, modern humans entered the scene in a very silent manner, and nobody can deny that up to this moment no scientist is utterly able to demonstrate, beyond any logical doubt, how and why we are here. It has been determined that the genus *Homo* appeared roughly during the late Pliocene, between two and three million years to present, and with the emergence of *Homo*, volume increments of the brain took place; from Lucy's brain (400 cubic centimeters) to modern human's (1,350 cubic centimeters). This noticeable increment forced to a more nutritious and steady diet, since the well developed brain consumes between 20 and 25 per cent of all nutrients ingested. Hence, with a very rudimentary diet supply it could not be possible for humans to develop and feed their hungry brains. Both events happened in a very coordinated evolutionary manner, one depending on the other, and both evolved extremely harmonized; we would not be what we are, otherwise.

We are a biological surprise, good or bad, heaven knows, but we are. I dare to say that in the record of species evolution, our emergence is an outstanding phenomenon, an unprecedented outcome, in such a dimension that despite the remarkable advances in many branches of Biology, molecular genetics included, a rotund answer is not available to explain our materialization as a species.

We are an old species, not because we have been a long time on Earth, that is not the point, we are old because we have the accumulated wisdom inherited from thousands of generations from our ancestors, probably from 130,000 years ago, or more. Yes, two skulls from Omo-Kibish area, in Ethiopia, were accurately dated as old as 130,000 years, and completely identifiable as *Homo sapiens*. Of course, we may not think of those ancient members of our species as we presently are, their *modus vivendi* was very different, hence, their culture and social development.

The knowledge received through the channels of language that poured the experience from millennia into every single new generation and ultimately through documents, along centuries of human written history, comprising what we call today the universal knowledge. Unfortunately, a great amount of such documents have been lost forever, as was the irremediable loss of the Bibliotheca of Alexandria, when Roman invaders destroyed and ransacked so a precious amount of the old world universal knowledge. A cultural loss that precluded future generations enjoy and acquire a treasure of ancient information and wisdom. The human rage destroyed the own human knowledge, accumulated for centuries.

Previous *Homo* species did not reach such accomplishments. They were a natural experiment, blind steps to forge the iron rungs to ascend in the evolutionary ladder they were caught in, with no consciousness of a future, not in terms of a long term goal; they were unable to foresee where they were walking to. Nothing in addition to stone tools, caves to dwell, and facing death-defying events day after day. Very simple minded, with no physical or psychological relax whatsoever.

It is supported by several researchers that ancestral humans lived between 50,000 to 90,000 years ago. In a period very near to 50,000 years to present, Africa underwent a difficult period of extreme dry climate, probably the reason why archeologists only detected ancestral human remains in the eastern part of the continent, since the rest of Africa was apparently uninhabited, since forests were reduced to patches and African grasslands were turned into dried prairies. In fact, nowadays western Africa is very difficult to scrutinize, since inclement weather with extremely high temperatures precludes camping for diggings, to some extent. In 1995, some remains assigned to an australo-pithecine, and dated around 3 million years, have been found west of the Rift Valley, which demonstrated that ancient homi-nids also occupied territories on West Africa. Obviously, this is an outstanding find, since it proves that not necessarily Eastern Africa was the cradle of humankind. What stay to be discerned is whether those remains are from individuals that moved there from the eastern area.

Very extreme calculations made by geneticists account for a population limit around 5,000 people, some 50,000 years to present. Though this figure may seem to be a very low mark, it was the foundation of the whole world population that expand-ed in a relatively few thousand years from that critical period on.

If this calculation is correct, the genetic implications that such a reduced ancestral human population had for the future development of *Homo sapiens,* was deterministic, regarding what we are and what we do today. First, 5,000 people is just a small-town-equivalent population, then **genetic drift**, the random process where the variant of a single gene may replace all other existing variants of that same gene, in a matter of just few gen-erations, may act without restraint to provide future generations with a unique, but dangerous gene pool. It is dangerous just be-cause the founder, being a small group, probably did not possess enough genetic variation to succeed in front of new pathogens and threatening ecological changes. In addition, new environ-mental changes could lead to misbalances in homeostasis, to-gether with all defective inherited traits concentrated in such a small group of humans. We, human beings, are so similar that our DNA all over the world seems to be the result of some virtual cloning process. Any group of chimps may show more genetic diversity than any geographic group of humans do. It is just be-cause all the people in the world share a very recent common ancestry, concentrated in a very small population. This extreme genetic similarity might be the direct result of our origin, from that small group of modern Africans that spread all over the world, carrying short-variable or poor DNA diversity. The genetic founder effect, from that small group, could be the reason for our poor DNA variability and undesirable genetic-load. This unpleasant genetic-load is more evident in certain human families that sadly carry a miserable genetic combination which is responsible for different and complex illness outcomes, from mental diseases to metabolic aberrations, sometimes converging in one single individual a disastrous inherited mixture of such ancestral dejected genes.

We are what we have in our DNA, plus the environment. Certain diseases do not appear due to the inappropriate environmental conditions for them to show up. On the other hand, that same individual may become a life-time ill person if placed in a different environment. Medical doctors know these cases as frequent issues. Asthma is, in general terms, a genetic based condition. Nevertheless, if the affected individual moves to another place, sometimes to another country, with different environmental surroundings, the respiratory condition will disappear, just because the environmental conditions drastically changed, and so the health condition. Unfortunately, not all genetic-based diseases behave the same. Defective genes may hurt several generations in a single family tree, as has been demonstrated in copious medical records ever since.

Almost certainly, that small group of African origin had been in the brink of extinction, perhaps more than once. Then again, let us take for granted the geneticists' calculations are correct, and the founders of our modern world population were just 5,000 individuals or so. Subtract children and elderly, which were not in reproductive conditions, plus the ones suffering different kinds of diseases; the active reproductive population size, effective population, was far below 5,000 individuals. From this small population, a migrating group started out of Africa.

Given the difficult living conditions, which produced severe environmental pressures, extinction was the expected outcome, but it did not happen. It is complicated to explain how they thrived, and their population grew to outnumber other human species and displaced them, in such a way that we have been the only living *Homo* species from 30,000 years to present, except for *Homo floresiensis*, as a rare relict population already discussed somewhere before in this book.

For ancestral humans, it appears that the biological laws, which are supposed to act under the above mentioned circumstances of the population, were silent, though all the conditions affecting that small population called for the application of said biological laws. What really happened that the principles of population genetics failed to act accordingly, allowing the precarious population size to overcome an irremissibly ongoing extinction? On an edge situation as it was, the worst was the expected extinction as a result, but...miraculously they overcame all possible difficulties for us to be here. I have been searching for some credible explanations in terms of population genetics to approach this unique occurrence; surmising, with or without tenable concepts, is all I have found and most of these statements do not approach the basis of population genetics at all.

We are, without a doubt, the victims of such a poor genetic endowment, which is the carrier of all the defects we suffer in our cheap DNA, including our ape-like aggressive behavior, which is out of context in a "theoretical" modern human being in a very modern society, as we consider ourselves. We all descend from a very small group of ancient humans, with a reduced genetic variability due to the undersized effective population. Consequently, we are not having the optimum variability that any species must have to thrive along the way, and the most evident outcome of all this is a defective genetic endowment. *Homo erectus* thrived for almost two million years on earth, it appears that they possessed a worthy DNA, probably very different from our poor one. I do not think we can emulate the long stance of *Homo erectus* and thrive as they did. We will not be here so long a time, our DNA does not allow such a prolonged existence as a species, though medical advances are trying to expand life expectancy; unfortunately, some diseases as cancer are increasingly rating. Medical advances are just trying to counteract the deleterious effect of our unfortunate DNA. If we stick to creation, it was imperfect. If we stick to evolution without creation, our evolution has been an erratic process to end up with an inherited ragged DNA, pitiable product after millions of years of biological evolution, with disastrous consequences for our quality of life, which in fact, is an extremely short spanned existence.

The homeostatic process is not enough to efficiently fight life-threatening diseases. If evolution, through natural selection, "created" the homeostasis as a disease-rejection mechanism to maintain the biological equilibrium, aimed to prevent fatal consequences when viruses, bacteria, and organic misbalances attack living things, then evolution fell too short to accomplish a finished job. Homeostasis is just a basic mechanism, a very basic one, intended just to keep a systemic equilibrium in the biological order. On the other hand, the real life-threatening diseases find no resistance in their course to damage our bodies, and finally they turn our existence into a few-decades-shrunk life span. Above and beyond, it becomes a suffering process that may lasts for several years before the fatal result, led by any disease, escort us to an end. Besides, our immunological system is an incomplete one, not strong and well structured enough to stoutly fight the real threatening disorders. Just think about flu virus, it can be devastating, even fatal to many persons. Our deprived immunological system is just a pitiable victim of such a virus, and many others.

We are unable to change the human genome, at least for the moment time, to set us apart once and for all, from our primitive ancestors. Nonetheless, that small group could thrive in spite of the adverse environmental conditions that prevailed thousands of years ago. We may deduce that their gene pool was at least composed with the "minimum genetic requirements."

though poor in its own nature, to survive the long journey of evolutionary path they faced.

It is, by all means, comprehensible that selective pressures were enormous, and, at any time during this process, the survival of the fittest was enforced *à outrance*. But even so, from a population genetics point of view, it is difficult to understand that such a small population, in fact an endangered species at the time, could survive when other human populations, much larger and genetically better, did not make it. Besides, they expanded their original limits very rapidly and populated the planet, replacing the existing *Homo* species, which were well adapted, sturdier, and accustomed to the environments where the ancestral humans, as newcomers, were not used to. This is an unprecedented event in the evolution of any species that formerly was at the edge of extinction...from rags to riches was the scenario for the play of those early *Homo sapiens*, our ancestors, from whom we inherited, in a major proportion, our modern genome.

This event is hard to explain based on natural selection, since such a process would push the minute population to extinction; there was not enough genetic variation to adequately respond to the environmental conditions that, at any time in the geological past, were naturally harsh for those primitive human beings, as has been stated by several authorities from different fields. Furthermore, what is more astonishing is the fact that they not only overcame the delicate and life threatening problem of an exiguous population; they populated the planet. Furthermore, in some periods, the earth population increased according to a geometric rate. Still, these consequences of genetic order, extremely well known, have not been treated *in extenso* by the experts, albeit many scientific articles and books have been printed on the subject of population genetics and evolution. In the circle of human knowledge, is common to avoid exactly those topics or issues that, at the light of orthodox science, are hard to handle due to a complete lack of available proofs, or strong evidences to start with.

As I expressed before in this book, not everybody dare to enter the "presumably unknown" with the audacity of avowals, which are not exactly on the ground of official or orthodox science, it is, inside the limits of nowadays conception of pure and genuine science. Any provocative statement may encounter the most recriminatory observations from the community of specialists. For them, it is better to leave the "unknown" on its place till more evidences may be available to play with. Any attempt to disclose other possibilities out of the orthodoxies firmly established, is something regrettable, though these could be just hypotheses to be proved in the future to come.

Besides, geneticists' studies on the mtDNA may confirm other important aspects and events in *Homo sapiens* ancient history. According to these studies, they have discovered certain mitochondrial lineages that are of a tremendous help in ascertaining events in our remote past. These authorities established that the lineages L1 and L2 from the mitochondrial DNA are presently restricted to Africa, while the female variants M and N of the L3 lineage left Africa in the postulated exodus of modern humans. On the other hand, studies on the Y chromosome (male exclusive inherited), cast results compatible with two distinct groups, the descendant carrying the mutation identified as M168 are found solely out of Africa; the other mutations of that gene are only found in the African continent. More striking is the fact that all genetic based studies point to the fact that the modern humans that migrated out of Africa were just a small group, as small as 170 people or so. It is simply incredible.

The reader may recall that, in previous paragraphs, it was explained that modern human population in Africa was calculated about 5,000 people, when extreme environmental conditions were onset about 50,000 years to present, and it was openly discussed the genetic implications that such a small population may carry for the future genetic diversity, which must be as ample as possible to provide a healthful assortment of genetic variants, to respond to unpredictable environmental changes; if such assortment, or gene pool, is poor in its own nature, then natural selection will not find the adequate genes to favor the ongoing evolution of the said population. Extinction could be a reasonable outcome to such impoverishment on that genetic endowment.

Well, what to say now if the calculation of the migrating group was just less than 200 individuals as stated before? Genetic drift, already explained in this chapter, must act severely upon this extra small migrating population, and the effects of gene substitution could be rampant, with the expected suppression of certain gene variants, which geneticists call alleles, and a total onset of other variants (alleles) up to the order of 100% in the population under genetic drift action; the smaller the population the stronger the effect of genetic drift on that population. It is such that, under the action of natural selection and genetic drift, a small population may change drastically in a relatively short period, then becoming different from the original ancestral population.

The biological problem is the fixation of alleles that compromise the health condition of the population, physically or mentally, or both. To start with a reduce endowment of genetic material is to condemn the descendants, from that genetically poor population, to a cruel inheritance as biological beings. With a poor genetic pool the members of such populations could survive only under a calamitous existence, trying to counteract the deficient DNA expressions inherited from their ancestors. Is this another version of the biblical punishment for the original sin?

At this point, the reader is perfectly identified with the fact that, the larger the genome the better the possibilities to thrive in spite of adverse conditions imposed by environmental changes. A rich genome offers the best possibilities for natural selection to choose the best genetic variants from a fair assortment, which was constructed along millions of years of evolution. But now, the inconceivable phenomenon is that, after millions of years of ascending evolution toward the upgraded *Homo* species that we are, what we have inherited is a poor, ridiculous, incomplete, and defective DNA. Is it worthwhile? If less than 200 modern people migrated out of Africa and, against all rational possibilities, moved following different courses, settled down, thrived, reproduced themselves in a very successful manner, and finally peopled the planet. Believe it or not, it is the reason for all the characteristics of our DNA.

Unfortunately, if the exodus of modern humans out of Africa was a reduced number of people, what we inherited is a very poor DNA...a decimated DNA. I have freely expressed before such consideration. Besides, chances are that deleterious mutations may appear along the way, to make it worse. That kind of mutations may act in many different ways, but always producing a threatening condition to the members of the population. Some of these mutations may impoverish health conditions in the individual carrying them, as is the case of all genetic-based illnesses that can be counted for hundreds, and that we suffer in modern times. Probably, most of these mutations were not present in our ancient ancestors; more recent ones are responsible of nowadays pathological conditions. Besides, under the evidences of so many bone and skeletal problems that humankind is suffering, according to world statistics, it is a basic sound reasoning to suspect that our skeletal structure is somehow under an inadequate biological level; it is not exactly what we need as modern humans.

Now, the route "out of Africa" for the modern human group is somehow under dispute, but certain facts, based on fossil records, shed some light on the route, I mean, the most probable route followed by that small group of moderns that decided to flee from Africa, though the continent was pretty large to guarantee plenty of food and shelter for the small communities that dwelled there thousands of years ago.

If the founder population, from where emigrants left, was situated in the area comprising present Sudan-Ethiopia-Kenya territory, the most probable route was through the Strait of Bab al-Mandab, which connects the Red Sea to the Gulf of Aden (see any Africa map as a reference). This strait was nearby the settlement of the original modern humans' population, or so believe. At that time, some 60,000 years to present, sea level was significantly below of what it is today. Probably, the strait was nothing but a shallow water pathway, which invited, as the easiest tour for them to go across and reach the Arabian Peninsula in its southern part, then kept on along the southern coast of India, to Indonesia. At that time Java, Borneo and Sumatra islands were connected, to form what is known as Sunda, already mentioned before.

Almost certainly, some of the emigrants that reached India continue the tour to the northwest, wandering through Iran and Turkey, and then arriving to European lands. This journey took some 16,000 to 18,000 years, according to certain calculations by specialists, and some crossroads inevitably occurred during the long term movement of those modern human emigrants, giving way to additional trails. Besides, along the route small groups settled down in different regions, as evidenced by fossils that have been found, while others continued and in turn settled down, according to their needs, in other geographical areas. Specialists have found remains of their presence that testify the ancient migration of our ancestors. Once in Europe, they spread over the years confronting Neanderthals, and finally pushed them to extinction.

Along all those thousand years, mutations would appear outlining differences among human populations in diverse regions, not only in the physical aspects of individuals, but in their trends to be healthy or stricken by different pathological conditions. Then again, based on that long journey under relentless environmental circumstances, acting upon mere migrating groups, which were facing unpredictable situations, arriving at new lands mostly or totally unknown for them, exposed to new and not well known edible stuff, it is completely astonishing that they survived and then, populated the world.

If we strictly stand on the side of scientific methodology and accepted theories, this survival event cannot be simply explained by population genetics, on the contrary, this discipline shall predict, based on simple calculations, the extinction or not thriving condition of such small and weak groups. Furthermore, if we take into account that the life span of our ancestors was short, mainly due to all kinds of contingencies on their way, it comes out that the reproductive life period was shortened as well, and the number of descendants shrank significantly. Much less can we evoke the action of natural selection, since selective forces steadily encountered a very poor gene pool to choose from; therefore, if any selection was made it was not the optimum. By all means, the selected genotypes, under the prevailing environmental circumstances, were not the best but mediocre ones.

In spite of all this population misbalances, *Homo sapiens* was able to completely replace Neanderthals and *Homo erectus*. Is there any scientific explanation compatible with 170 migrating people facing all kinds of eventualities and after all those calamities a happy ending was disclosed, in such a way that all other *Homo* species disappeared and we are here? Besides, those invading modern humans by no means were as many as the established Neanderthals. Above all, they were newcomers. Neanderthals were adroit in managing the basic resources of those lands, as their permanence for much more than 100,000 years proved. As I said before, guessing is the only way out in such a complicated and non-natural selection dependent issue, and that is exactly what I have read in several articles and books on this matter, a kind of guessing, as there is not a solid grasp on genuine science to convincingly explain such an odd event.

Hunting-gathering had been the only sustaining-life activity known to humans for an immense period, and then in a crucial short-elapsed time people started to behave different, probably about 11,600 years ago when the world climate entered and inter-ice era, that still endures today. Previously, a warming period that is calculated from 14,500 to 20,000 years ago, the planet enjoyed that warming phase, but suddenly an extreme cold period covered Eurasia, which sent humans again to a difficult acclimatization and survival conditions, hard for them to efficiently face this sudden climatological severe situation. It lasted for about 1,350 years or so, according to estimations on climate for that time. Surprisingly, this rough period abruptly ended, following a sudden pattern as when started.

As this last glacial extreme or Glacial Maximum ended, as many specialists describe it, a new turn in human living style began to be established for the first time in the long history of *Homo* species. Somewhere, in the near east, an incipient new kind of social-sharing human group began to flourish with an unprecedented vision of group, not the wandering group as it had been before, but one that settled down for the first time to live upon another conception of sharing, protection and commitment. How this exactly happened? I do not think the answer is simple or even possible, though some smart guessing may lead to reasonable explanations that may spotlight on the logical events and situations, which finally, made possible for those ancient groups to settle down. Authorities conveyed that this was a true revolution in ancient human social relations. It was the starting point for all happenings that were yet to come.

It is evident that, since the Paleolithic Age began 500,000 years ago and mainly characterized with primitive hunter-gatherers as the best social relation, the human settlements became the most extraordinary event to impulse community relations to convert human groups into societies. During the whole Paleolithic many and profound climatic changes took place; unquestionably, all these events molded humans into the modern mind and body to develop new technologies to cope with a settle-living style. All this happened in the last period of the Paleolithic Age.

According to geologists and other authorities, at the end of the last ice age between 6,000 to 10,000 years ago weather changed considerably, producing among other changes the rising of sea level; salinity and temperature became different. Furthermore, short before the end of that last ice age, the extinction of large mammals occurred covering the main continents, Europe, Australia, North America, and Asia. It is believed that this mass extinction impacted human feeding habits in a severe manner, urging those ancient humans to look for alternate food sources, being edible plants a major part of it, or at least, an important part of their diet to rapidly compensate the scarcity of animal protein.

It is reasonable to judge these circumstances as an advocate for the posterior development of agriculture some 7,000 years ago, which marked the end of the Paleolithic Age. Though it is somehow uncertain, all evidences point to agriculture as the plunge to domestication of animals, mainly those associated with hunting activities, the dog, and all that could represent a source of protein what was provided by the extinct large mammals, already mentioned, which occurred as the first Holocene mass extinction.

Back to the starting up of settlements, different aspects could be considered, perhaps all of them acting as a complex condition on those primitive humans day-by-day experience. A possible reason, as pertinent to such decision as "to stay instead of leaving," probably was based on the most important aspect of daily life for those primitive small communities, it is, food. If the abundance of small game, fishing, and edible plant resources on a certain area were enough to sustain the group, it is likely that the determination to settle down was not exactly a "once and for all" decision. It is plausible to assume that it started as a fairly temporary action based on the said abundance. As time went by, and everything was going right, and the food was obtained plentifully together with a source of drinking water nearby, the temporary settlement was extended in time, and they learned how to go around and get what they needed, and then return home where a secure shelter was always there. At the same time, as it became a more cohesive group the protection was incremented, and the security for the whole small community increased as well. They learned, probably in the hard way, the benefits of common living, sharing everything in a more convenient way than as hunter-gatherers did rambling around, with poor cohesion. Perhaps the perils of attacks from outsiders were increased; they had to learn how to organize a better defense against aggressors. It was the primordial embryo of the human community. Attackers, criminals, and opportunistic marauders were always a menace in the nearby surroundings of primitive settlements, as they are today.

All of the above are nothing but smart guessing, since the fact that change *Homo sapiens* way of thinking and metamorphosed a hunter-gatherer into a modern settler in a relatively short period, could be the result of a very rapid course of selection, drastically and qualitatively changing the structural thinking process, plunging those ancient groups on the trail of the most revolutionary achievement of humankind, the one that utterly turned wanderers into very modern human beings, up in the ascending ladder of civilization. But, what kind of trait was selected to produce such a drastic change throughout human behavior? As much as have been written on this topic, no convincing explanations are at hand. Morphological and physiological changes, or any other trait, are easier to explain and understand than an extreme twist in behavioral patterns, which according to experts require very profound changes of neurological origin, which are by no means simple modifications, not to mention that this behavioral change occurred in a relatively short period, as was stated before. The other alternate explanation is the action of inexplicable or unknown natural forces that led all *Homo* species along a journey of about 500,000 years, up to the point of complete humanness, at least, in terms of technology and social communities.

Besides, all those changes that rapidly impulsed the biological emergence of modern humans, may be considered from two possible stand points: a)mutations that led to the accomplishments of unprecedented tasks, or b)a dormant genetic endowment already existed, which came activated because of the proper environmental conditions or both, in a very coordinated manner acting together as an extraordinary biological gear, where every part of the system was placed at the exact time to lead the outcome of a complete human being, capable of accelerating the cultural, technological, and social progress for the future to come. No possible biological accomplishments could be obtained if they are not supported on a genetic basis. The genome is the reservoir of all possible changes and modifications. If there is not the proper information in the genome, populations and individuals cannot biologically evolve, which means that not only from a physiological or anatomical point of view is the evolution possible, but psychological accomplishments have to be performed as well. From this point of view, it is somehow difficult to understand how natural selection could act so quickly to fully develop *Homo sapiens* up to our days.

Science is not in the position of calculating what the probability is for the emergence of humans, through natural selection and evolution, exclusively. It is almost an impossible calculation, simply because many variables must be taken into account, and most of them are unknown. Nevertheless, if we consider the 500,000 years of human evolution, the different hominid species that comprise the human lineage from the very beginning, the fact that probably several hominid species had been in the brink of extinction before further evolution, which would lead to the termination of the human evolutionary process, the different attempts of human models as Neanderthals, *Homo floresiensis*, and why not, *Homo erectus* to thrive and evolve to higher degrees of perfection, but they did not, and finally, the complete sweep-off of all those human models and the supremacy of *Homo sapiens*, as the only and unique human species on planet Earth may represent a very small probability, probably one in several millions. Recently, March 2010, a finger bone was collected in a Siberian cave. According to scholars, this remains belonged to a hominid that probably lived alongside *Homo sapiens* and Neanderthals, some 40,000 years ago. Besides, DNA studies put it distantly related to both modern humans and Neanderthals. It seems to be a new species not found before, which makes evident the co-existence of three different human species in that area. Studies continue to determine the nuclear DNA sequence, which is more complete, and eventually will provide a tremendous amount of information on this new hominid. Once again, at this point we are everyday more uncertain how many possible human forms existed in the past and went extinct.

Science is not either in the position of accepting any other alternate possibility but "evolution based on the previous selective work of natural selection." Even so, when science is facing very delicate accomplishments in the evolutionary order, and it sounds like natural selection is not enough to fully explain the event under analysis, there is no other plausible way to approach such occurrence but to expand our criteria no matter how far. Science cannot limit itself based on official policies and narrow conventionalities; science has to outreach the limits of its own epoch. Darwin did it, and now we have the most extraordinary body of theory to explain how the evolutionary process works, to some extent. We should not impose limits to scientific thoughts and development; we are pushing back the advancement of science, otherwise.

No matter how daring a proposal might be, if not at present, probably in times to come the answer or the proof, will be held. Besides, if we stick unconditionally to the criterion that natural selection is an immovable fact and that any evolutionary event in biology has to be explained by its action, then we are denying the principle of development, since natural selection as a theory is subjected to modifications and interpretations, according to the advancement of human knowledge. If we are consequent evolutionists, we cannot crave for static theories, no matter how useful they are or have been. Perhaps not all evolutionary events are completely and satisfactorily explained through the principles of natural selection. We cannot make the same awful mistakes that Marxism followers make when they impose any analysis, exclusively, under the principles of the *dialectic materialism,* as the only reliable theory to conduct genuine scientific methods.

The projection of the scientific mind must not have any academic limit. The scientist, as a free thinking individual, is opened to project his/her hypotheses and theories fearless to the future, or back to the past, as is the case when it is needed to explain difficult events from remote times. These projections must be stated regardless of the prevailing orthodox policy, if they don't, they are imposing and sustaining, without knowing, the shadow of an unexpected neo-inquisition era, otherwise.

At the time settlements were taking place probably as a sporadic event, language was at a peak of development, which was a condition to expand and organize the community at its very beginning. It is almost unattainable to think of the most basic organization if language was not present and well developed. Perhaps, and probably it was, the new conditions of settlement-living generated new vocables that were implemented based on the needs of novel circumstances, not faced before by ancient human groups. Many linguists agree that new living conditions behave like a trigger to promote new words and expressions; the modern history of humanity is a glowing example of such radical, and in many ways innovative, new turns of languages all around the world. English language, itself, is a paradigmatic example of such principles.

As settlements were dispersed, and every single settled group evolved with varying needs and environmental conditions, it is not reasonable to talk about one single and inflexible pattern of language development. This is exactly the reason why diversification in language structures took place. It is just very similar to nowadays cases in which every single modern family has certain way of saying, and use specific expressions that other families never use or infrequently use. These expressions pass from one generation to the next as inherited through the line of descendants, and certainly, with new generations new expressions and refrains are established *de novo*. We can witness exactly a modern model in families and villages today that may resemble, to some extend, what was the onset of different language modes from afar of the written human history.

More Ancient Kin

During the first part of this chapter an account of *Homo* species has been the main topic, highlighting the aspects that, according to the scope of the book, are of the main interest to assemble, in a non perfect way, the major paths in our evolution, considering the most recent species, which may be judged as humans; no doubt about it. Again, all preceding pages of the chapter grouped together those species and their relevant characteristics. I consider necessary to leap back, trying to see, or imagine, what the predecessors of *Homo* can tell us.

The Sadiman volcano, in east Africa, was the direct helper, through its eruption and deposition of ashes that made possible for Leakey's team to find the footprints of perfect upright walking creatures at Laetoli. Those ashes preserved the footprints of two very ancient predecessors of our species that around 3,600,000 years ago walked side by side leaving their footprints as indelible marks for us, modern humans, to know about them and their existence.

Scientists agreed that those two individuals were members of the *Australopithecus afarensis* species, real ape-human creatures that were perfect walkers, and what is more relevant, their footprints are almost indistinguishable from ours; so human were their feet. It is not difficult to assume that as they lived from 3 to 4 million years ago, our bipedal condition was already developed, at least, at that time. Furthermore, a presumable ancestor of *Australopithecus afarensis* (*Australopithecus anamensis*) has been known by some bones found in northern Kenya. These remains were dated as old as 4,300,000 years to present, and they show unequivocal upright walking features, since leg joints are similar to those of *Australopithecus afarensis*. If this relationship is true, and it seems to be, there is a direct relation ancestor-descendant comprising both species. However, there is a consensus that there is only one species, *Australopithecus afarensis*, the other, is an artificial and arbitrary designation; it is nothing but an evolving lineage through geological time with the morphological variations normally expected in an evolutionary process.

The reader may refer to Fig. 8 to recall what the position of *Australopithecus afarensis* might be regarding the phylogeny of humans. It is worth to mention that this species, according to its long arms, switched from walking to moving in the branches of trees, using both locomotion styles in its favor. Nonetheless, other bone characteristics might be remnants of ancestral forms, since not all anatomical features evolve at the same rate, as has been clearly stated by scholars. Based on this, bipedalism in *Australopithecus afarensis* could be the major locomotion style. Besides the genuine footprints, pelvis and limb bones are typical of a bipedal species, sporadically climbing trees in search for fruits, nuts, and other edible parts of certain plants. This assumption is supported by the long curved fingers and long forearms. Besides, according to certain finds, related to paleoenvironmental studies seriously performed, it comes out that there is enough evidence to believe that bipedalism evolved in the forest, long before that hominids ventured into grasslands. Long arms and short legs in *Australopithecines* may account for the fact that albeit they were walking creatures, their arms kept the dimensions and characteristics typical of forest dwellers, which were related to tree climbing. In this way, the reduction in the length of the arms was a subsequent accomplished feature in the evolution of the human shape.

It is considered that *Australopithecus afarensis* was a real thriving species for more than a million years. At present, it is not a simple task to determine how well adapted was this species to its environment in such aspects as protection of the group, cohesion of the group, and moving in search for food and shelter. But one thing is very clear, they stayed on earth for such a long period that its adaptability, in general terms, is out of the question. It is believed that they mainly lived on vegetarian diet, though it is completely possible that small animals, of any kind, could be included in their diets, as well as are included in chimps' diet, as a supplement of proteins.

More recent fossils of the same species made possible to calculate their tallness, which ranges from over 5 feet to a little more than 3 feet. This span has been interpreted by specialists as an indication of an evident dimorphism between sexes, as is the rule in apes, like gorillas and chimps. Their perfect bipedal condition put them on the way to humans. Their long arms, rather short legs, pronounced mouth-nasal portion set them close to apes. In fact, we may imagine, as many scientific reconstructions and versions have been achieved in museums, they were exactly ape-human creatures, like a rather well cast hybrid-morph of both species, only possible through a precise evolutionary process that leaving the apes behind, started to shape the future human beings that we are today. At this point, we cannot forget that any evolutionary step toward humanness was only possible if the genome of those creatures were endowed with the precise genetic information to evolve toward superior forms, not only in terms of adaptation, but as an accurate evolutionary path, heading higher levels to become extraordinary biological beings, not suspected before in any former species that populated planet earth during the preceding millions of years.

Studies comparing skeletal elements proved that *Australopithecus afarensis* show similar sexual dimorphism in variables, from femur bones, as gorillas. On the other hand, if humerus bones are considered, they show similar to humans. Finally, canine variables are similar to chimps regarding sexual dimorphism. These shared similarities evidence that skeletal variables, concerning sexual dimorphism, cannot be completely ascribed to humans or apes, but rather show themselves as a mosaic to shape an ape-human species; a very primitive hominid.

Donald C. Johanson described Australopithecus afarensis in 1974, and the skeleton found, pertaining to a female, was named Lucy. It has been an anthropological icon ever since and probably represents the most important find during the twentieth century, regarding human evolution. Its discovery proved, for the first time, that the anatomical structure of our feet was already present in Lucy's. Her skeleton was dated 3,200,000 years old.

Australopithecus afarensis was much more than a chimp, much less than a human. There is something very interesting in the geological relation between *Australopithecus afarensis* and the appearance of *Homo.* There is an unsolved gap between both groups of species. *Australopithecus afarensis* disappeared, then, about 200,000 years later *Homo* species appeared in the geological records. Several specialists have stated this as an unsolved mystery. Nobody can tell with any degree of certainty what really happened in that one fifth of a million years. How *Australopithecus afarensis* evolved into *Homo?* How many "intermediate" human-like forms did emerge during the great gap? How did *Homo* come to planet Earth scene? Besides, in the evolutionary framework of the multifaceted *Homo* species, a lapse of only 200,000 years is merely a blink in geological time; to short to run from *Australopithecus* to *Homo,* a complete and genuine human being. What really happened to accomplish such an evolutionary leap in such a short period? On the other hand, it has been tradi-tionally considered *Australopithecus africanus* as an intermediate species between *Homo* and *Australopithecus afarensis*, hence as a direct Homo's ancestor, but strong evidences published by White and Johanson in 1981 proved *A. africanus* evolved toward, or branched to *Australopithecus robustus* not toward *Homo*. In Fig. 8, a question mark denotes the improbable position of *Australopithe-cus africanus* as a direct *Homo* ancestor. Besides, it is important to express that *Australopithecus afarensis* was not the immediate antecessor of the genus *Homo*; it had three descendant lineages, or so believe, the last one led to *Homo*.

Furthermore, Darwin for the first time stated the idea of linear-descent with no branching, known as anagenesis. This could be the case for australopithecines' evolution; more fossil materials are necessary before any serious statement could be mentioned in that direction. If anagenesis is the evolutionary process that took place in the australopithecine lineage, then it sounds plausible to infer that such an evolutionary pattern could speed the evolution of hominids toward the genus *Homo*. Without branching all the evolutionary process was following a quick-ladder of results; no branching-crossroads. If this hypoth-esis could be proved, then another question may arise: why ana-genesis led the process during the last steps of human evolution while in previous periods branching was the rule?

Unfortunately, as has been expressed before, many fossils have been lost in the sands of time. Hence, many intermediate and important human species shall never come out from their earthly resting places, to become a fossil of help in the realm of the relentless interpretation of human evolutionary path, not only for the difficult task that implies to find fossils, certainly fortuitous events, but for the disappointing fact as to have the remains of unknown human species turned into soil forever. Nonetheless, why is that no fossil records have been found to attest archaic gorillas and chimpanzees as we have for *Homo* lin-eage? Specialists openly declare that this mystery also remains unsolved. The so awaited common ancestor of chimps and *Homo* is yet to show-up. Surprisingly, an unexpected boom occurred in 2009. At the beginning of October that same year a spectacular article was published on the complete study of *Ardipithecus ramidus*, whose remains were collected seventeen years before. During all those years an extremely careful lab work was accomplished to obtain 125 bone pieces, which were embedded in the rock matrix, in such a way that very precise procedures were applied to recover, in pristine conditions, those remains.

This new species was dated 4,400,000 years old, and astonishingly they were upright walkers. The skeleton named "Ardi" shows, according to some paleoanthropologists, a strong evidence that the last common ancestor we share with chimps was not necessarily a chimp-like species or something close to human in appearance. Ardi is a collage of modern and ancestral features, which in turn, may be an anthropological projection or time-mirror of what the common ancestor we share with chimps might look like.

Something real new in relation to Ardi, is that, though joints and bones in legs and hands, including fingers, feet, and pelvis, clearly indicate a bipedal motion, quadrupedal was the style when moving in the trees, nothing similar to modern apes. This last characteristic evidenced by the large toe in diverging position as seen in chimps and gorillas. This adds the detail that albeit Ardi walked upright on the ground, was not as efficiently as *Australopithecus afarensis*. In fact, scientists agree that a decisive criterion to determine a true bipedal mode and not a quadruped animal lies on the cross-section of the neck of the femur bone, since during bipedal locomotion the head of the femur is able to transfer the weight load of the body (legs not included) directly to the lower limbs.

Together with Ardi, bones of other 36 individuals were collected as well. Together they found thousands of animal fossils, which made possible a very complete picture of Ardi's surroundings and ecology, which turned out to be woodlands. Then, a walking creature living full time in woodlands completely demolishes the "out of the forest" theory to explain bipedalism. If Ardi, which was a female according to measurements and exact determination of anatomical features, lived 4,400,000 years ago and was a bipedal species, then it is not adventurous to advance that bipedal condition was probably established long before, perhaps in the common ancestor itself. Besides, if Ardi's finger bones are not consistent with a knuckle-walking manner, then, if scholars are right, perhaps our ancestor not exactly went through such locomotion style, either.

Other human-like traits found in Ardi are the canines, which are small like in humans; males and females are very closed in tallness, which is another human characteristic.

Nonetheless, the possibility that *Ardipithecus ramidus* was in the human evolutionary line, is not unanimously accepted, since another possible outcome might be a dead end species, in the long evolutionary path of the unknown labyrinth that shape our course on planet earth. The authors proposed the emergence of *Ardipithecus* from the last common ancestor that humans shared with chimpanzees. First of all, it is scientifically mandatory to prove that *Ardipithecus* is an ape-like form in the pre-hominid line, since *Ardipithecus* is not included in the *Homo* genus and may only be another form among many others that existed and became extinct, just as mere splitting branches of the evolutionary phenomenon. We cannot take lightly that Ardi's feet are in straight connection with chimps and gorillas. Besides, its quadruped style sends Ardi back to more primitive monkey species. This collage of biological traits present in *Ardipithecus ramidus* may indicate a not well stabilized species, but a mosaic of alternate features, as many other species could be in the long remote past. Though really important, this new species is nothing but another piece in the big evolutionary puzzle.

Every new find is just a piece of the puzzle, which is not possible to complete in full details. With every new piece, new hypotheses emerge and open discussions gain the stage of work-shops, symposiums and lectures. Scholars are honestly engaged in the fruitful study of any remains that may shed some, some-times insignificant light on any aspect of our incomplete history as a species on our planet. Ardi is doubtless a very important find, but more material is necessary to assess unequivocally the exact evolutionary position of *Ardipithecus* in the intricate and zigzagging trail of our evolving species.

Different authorities have described several Australopith-ecine species. At present they add to seven different species; some of these have been described based only in some few in-complete bones. The point is that each single specialist believes his/her finds are different from the preceding described species and then, a new species is named according to diverse criteria and opinions hold by that particular researcher. Not only in an-thropological grounds have splits of species occurred. Along the history of systematic zoology, in general, numerous populations have been considered as different species, when in fact only one species with a wide ecological variation is the issue. Then again, in systematic zoology some specialists propose new names for species already described in the past, now based on recent DNA studies. To scientifically ascertain a true and different species, DNA is an invaluable help, but cannot be the only and exclusive tool to determine a species phylogenetic position. Any other con-tribution, as morphology, ethology, ecological data, etc. ought to be considered as very important, too; those specialists that constraint the diagnosis exclusively to DNA results, are reduc-ing and simplifying a multi-factor phenomenon to a single scope vision. In systematic zoology specialists are dealing with whole animals, alive, and in their natural habitats. There is nothing to guess, or suppose. On the other hand, paleoanthropologists are dealing with remains, in many cases with a small bunch of cracked bones, incomplete skulls, and scatter teeth. Evidently, chances are that smart guessing fails in determining the accu-rate position of a human or pre-human species in the evolution-ary context, with so scarce material available.

If systematic zoologists have to be very careful about nam-ing new species, anthropologists have to be so much more. I al-ready mentioned that life is variability, hence, careful assessment on any biological material is mandatory; the specialists could be complicating the evolutionary stage, otherwise. Neanderthals's remains prove there is an evident variability in skeletal struc-tures. It is well known today that Neanderthals from warmer regions were not so stocky, rather more lightly bone structures characterized them all. Species variations are readily observed when the distributional area of a certain species is large enough. This is another principle in population genetics that has to be considered when dealing with any species, no matter if they are humans or not.

The experiment is simple; just take a careful look at differ-ent people in any crowded place. You will find people with long faces, broad heads, projecting jaws (prognathism), with big and small noses, with a little depressed frontal bones, etc. Particu-larly take a careful look at the face portion from the base of the nose to the bottom of the chin. Some modern humans bear that portion of their faces longer and broader than their forehead. It is enticing to imagine a massive jaw bone in such individu-als. Yes, usually those people bear broader jaw bones than any regular member of the population, but they are as modern as the rest. All these are normal variations within our species. So, there is nothing weird that many fossil variations belong to the same species, and there is no ground enough to consider differ-ent species, and name them apart. It is just the natural variation or intra-population variation, readily expected. Precisely, based on those variations natural selection acts, and selects according to environmental circumstances, favoring certain traits and dis-qualifying others.

A genuine biologist cannot be fooled by variations; if so, he/she might be ignoring the top principle of biology, variation. How could they support natural selection on one hand and in the other ignore variation, which is itself the raw material for natural selection? Inconceivably they take variations to support their species-split procedures. In the systematic jargon these specialists are known as "splitters." This does not mean that factually many hominid species did not exist along the evolutionary road. In previous paragraphs, it was openly affirmed our ignorance on many species, which fossil records will never be available. And worst, we cannot imagine or roughly calculate how many species have been lost during the evolutionary process of humanization, not even how many dead ends were reached, which biological outcome never was.

It is astonishing the long, extremely long way of human evolution comprising an unknown amount of species that unfortunately could not leave their signature on the geological records. It seems like, if one particular attempt failed throughout humanness, another was on the way to correct the mistakes evolution made with the previous forms, in the incommensurable long parade of humanoid forms, technically known as hominids. It is evident, from that point of view, that the onset of humans was not an easy biological process; on the contrary, the long evolutionary journey was a tortuous road, which led the relentless process of becoming humans. No matter how hard, painful, erratic, biologically mistaken, and difficult to achieve the process was, humans ought to be here. It was a stubborn course of action that faced all imaginable stumbles on its way, but we are here.

Many scientists agreed that it is a unique phenomenon, difficult to believe it could become a success with so many obstacles on its way. If it is certainly one of a kind biological event, then…, are we anything like "a miracle"? Some specialists rather named this extraordinary process as a product of good fortune. For heaven sake! In the natural world, under uncountable ecological pressures, "good fortune" is not a scientific expression to apply on this extraordinary process that ended up with the emergence of modern humans, and it continues to higher stages, though imprecisely understood and difficult to forecast by sociologists, psychologists, paleoanthropologists, and other related specialists all the world over.

Perhaps, the most outstanding event within the event of intellectual and creativity boom was that all this happened almost simultaneously, all over the distributional area of *Homo sapiens* some 40,000 years ago. How this could happen if there were no means of communication among *Homo sapiens'* populations, far apart thousands of miles. What was in the neural systems of all human populations at the time to explain the outburst of unprecedented artistic creativeness, abstract thinking, and intellectual skills not evidenced before in such a finished and delicate performance?

Sadly, our ancestral history is highly fragmented over millennia and almost unknown when we enter the anthropological dimness of millions of years…that incommensurable darkness of time. We have to be almost pleased in approaching our pre-history as we approach an incomplete 30,000 piece-puzzle, which we received from an unknown printing procedure, with blur colors, due to the time elapsed, hardly snapping-on to get the completely smooth picture of the work of art and then, to complete the frustrating reconstruction…missing pieces are all over the frame.

Recommended Bibliography

Anton, S., Leonard, W. R., Robertson, M. L. 2002. An Ecomorphological Model of the Initial Hominid Dispersal from Africa. J. of Human Evolution 43: 773-85

Arsuaga, J. L., Martinez, I., Lorenzo, C., Gracia, A., Munoz, A., Alonso, O., Gallego, J. 1999. The Human Cranial Remains from Gran Dolina Lower Pleistocene site (Sierra de Atapuerca Spain) J. of Human Evolution 37: 431-57

Bandelt, Hans-Jurgen, Macaulay, V. and Richards, M. 2006. Human Mitochondrial DNA and the Evolution of *Homo sapiens*. Springer-Verlag. Berlin. 271 pp.

Bickerton, D. 1992. Language and Species. University of Chicago Press. 305 pp.

Boaz, Noel, T. and Ciochon, Russell, L. 2004. Dragon Bone Hill. An Ice-Age Saga of *Homo erectus*. Oxford University Press. N.Y. 232 pp.

Brown, P., Sutikna, T., Morwood, M. J., Soejono, R. P. Jatmiko, E., Saptomo, W., Awe Due, Rokus. 2004. A New Small-Bodied Hominin from the Late Pleistocene of Flores, Indonesia. Nature 431: 1055-61.

Coqueugniot, H., Hublin, J. J. 2004. Early Brain Growth in *Homo erectus* and Implications for Cognitive Ability. Nature 431: 299-302

Corrucini, R. S. and McHenry, H. M. 2001. Knuckle-Walking Hominid Ancestors. J. of Human Evolution. 40: 507-11

Enard, W. 2002. Molecular Evolution of FOXP2, a Gene Involved in Speech and Language. Nature 418: 869-872

Falk, D., Hildebolt, C. 2005. The Brain of LB1 *Homo floresiensis*. Science 308:242-245

Filler, A. G. 2007. The Upright Ape-a new origin of the species. New Page Books. N.J. 288 pp.

Finkel, M. The Hadza. 2009. National Geographic. 216(6): 94-118

Foley, R. A. and Lewin, R. 2004. Principles of Human Evolution. Wiley-Blackwell. Second Edition. 576 pp.

Fragan, B. 2010. When We Met Them. Discover 5-7. September.

Gwin, P. 2008. Lost Tribes of the Green Sahara. National Geographic. 214(3): 127-143

Hublin, J. J. 2009. The Origins of Neandertals. Edit. Richard G. Klein. Stanford Univ., California. PNAS. 106: 16022-16027.

Jablonski, N. G., Chaplin, G. 2000. The Evolution of Human Skin Coloration. J. of Human Evolution 39: 57-106

Johanson, Donald, C. 1996. Face to Face with Lucy's Family. National Geographic. 189:96-117.

Johanson, Donald, C. and Wong, Kate. 2009. Lucy's Legacy. The Quest for Human Origins. Harmony Books. N.Y. First Edition. 309 pp.

Kimbel, W.H., Lockwood, C. A., Ward, C. V., Leaky, M. G., Rak, Y., Johanson, D. C. 2006. Was *Australopithecus anamensis* ancestral to *A. afarensis*? A case of Anagenesis in the Hominin Fossil Record. Journal of Human Evolution. 51:134-52

Kline, R. G. 2009. The Human Career: Human Biological and Cultural Origins. University of Chicago Press. Third Edition. Chicago. 1024 pp.

Krings, M. 1997. Neanderthal DNA Sequences and the Origin of Modern Humans. Cell 90: 19-30

LeBlanc, S. A. and Register, K. E. 2004. Constant Battles. St. Martin's Griffin. Publisher. 256 pp.

Lovejoy, C. O. 2009. Reexamining Human Origins in Light of *Ardipithecus ramidus*. Science. 326(74): 74e1-74e8

McBrearty, S., Brooks, A. S. 2000. The Revolution that Wasn't: a New Interpretation of the Origin of Modern Human Behavior. J. of Human Evolution 39: 453-563

McKie, Robin. 2000. Dawn of Man. The Story of Human Evolution. Dorling Kindersley Pub. Inc. N.Y. First American Edition. 216 pp.

Morwood, M. J. 2005. Further Evidence for Small-Bodied Hominins from the Late Pleistocene of Flores, Indonesia. Nature 437: 1012-17

Neimark, J. 2010. Who's in our Genes? Discover 78-79. September.

Relethfor, J. H. 2008. Genetic Evidence and the Modern Human Origins Debate. Heredity. 100: 555-563

Rightmire, G. P. 1990. The Evolution of *Homo erectus*: Comparative Anatomical Studies of an Extinct Human Species. Cambridge. Cambridge Univ. Press. 276 pp.

Rightmire, G. P. 1998. Human Evolution in the Middle Pleistocene: The role of *Homo heidelbergensis*. Evolutionary Anthropology. 6: 218-27

Rightmire, G. P. 2001. Patterns of Hominid Evolution and Dispersal in the Middle Pleistocene. Quaternary Int. 75: 77-84

Serre, D. 2004. No Evidence of Neanderthal mtDNA Contribution to Early Modern Humans. Public Library of Science. Biology. 2: 1-5.

Shreeve, J. 1995. The Neandertal Enigma. Solving the Mystery of Modern Human Origins. William Morrow and Company, Inc. N.Y. 369 pp

Shreeve, J. 2010. The Evolutionary Road. National Geographic. 218 (1): 35-67

Sloan, Ch. P. 2010. A Revealing Relative. National Geographic. 217 (6): page 24.

Stanford, C. B. 2003. Upright: the Evolutionary Key to Becoming Human. Boston. Houghton Mifflin Harcourt. 224 pp.

Steegmann, A. T., Cerny, F. J., Holliday, T. W. 2002. Neandertal Cold Adaptation: physiological and energetic factors. Amer. J. of Human Biol. 14: 566-583

Stock, G. 2002. Redesigning Humans: Our Inevitable Genetic Future. Boston. Houghton Mifflin Harcourt. 288 pp.

Stringer, Ch., and McKie, R. 1998. African Exodus: The Origins of Modern Humanity. Owl Books. 272 pp.

Swisher,III, C. C., Rink, W. J., Anton, S. C., Schwarcz, H. P., Curtis, G. H., Suprijo, A. 1996. Latest *Homo erectus* in

Java: Potential Contemporaneity with *Homo sapiens* in Southeast Asia. Science 274: 1870-74

Tattersall, I. and Schwartz, J. 1999. Hominids and Hybrids: The Place of Neandertals in Human Evolution. Proceedings of the National Academy of Science. 96: 7117-19

Tattersall, I. and Schwartz, J. 2001. Extinct Humans. N.Y. Nevraumont Pub. Co. 256 pp.

Templeton, A. R. 2002. Out of Africa Again and Again. Nature 416 45-51

Templeton, A. R. 2005. Haplotype Trees and Modern Human Origins. 2005. Yrbk. Phys. Anthropology 48: 33-59

Templeton A. R. 2007. Genetics and Recent Human Evolution. Evolution 61: 1507-1519

Tzedakis, P. C., Hughen, K. A., Cacho, I., Harvati, K. 2007. Placing Late Neanderthals in a Climatic Context. Nature 449: 206-208.

Ward, C. V. and Leaky, M. G. 2001. Morphology of Australopithecus anamensis from Kanapoi and Allia Bay, Kenya. J. of Human Evolution 41: 255-368

Wade, Nicholas. 2007. Before the Dawn. Recovering the Lost History of Our Ancestors. Penguin Books. N.Y. 314 pp.

White, T. D., Johanson, D. C., Kimbel, W. H. 1981. *Australopithecus africanus*: its Phyletic Position Reconsidered. South African Journal of Science. 77: 445-70.

White, T. D., Asfaw, B. 2003. Pleistocene *Homo sapiens* from Middle Awash, Ethiopia. Nature 423: 742-747.

White, T. D., Asfaw, B., Beyene, Y., Haile-Selassie, Y., Lovejoy, C. O., Suwa, G., WoldeGabriel, G. 2009. *Ardipithecus ramidus* and the Paleobiology of Early Hominids. Science. 326 (64): 75-86

Wong, Kate. 2003. An Ancestor to Call Our Own. Scientific American. Jan. 54-63.

Zimmer, Carl. 2005. Smithsonian Intimate Guide to Human Origins. Smithsonian Books. Madison Press Books. Toronto. Canada. First Edition. 176 pp.

Chapter 4

Chimpanzees and Bonobos. The Living Kin

Not less than six million years ago, a yet unknown hominid form split from a common ancestor shared with chimpanzees. This is the most accepted event regarding the beginning of the human lineage. For most paleoanthropologists, we are only and nothing but chimpanzees' cousins. In fact, it depends if the common ancestor gave way to a *Homo* species. If it was the real event, then chimpanzees are not merely our cousins, they are our brothers; we share a true common ancestor. This concept exactly pertains to the present chapter. Perhaps, for many people in our world, this affirmation is too much. Then again, remember that we share with chimps more than 98% of our DNA. This is not a matter of believing, this is a scientific fact, well known since many years ago. Despite the divergence between chimps and humans occurred millions of years ago, the similarity between both genomes is remarkable. At present time, the Chimpanzee Genome Project is another scientific fact, which is providing a tremendous amount of information on genetic grounds. This project started up in 2005, and is conducted by well-known scholars from different top of the line institutions in the United States.

Human and chimpanzee's genome comprises close to 40 million single base pair substitutions (see chapter 1 as a reference for DNA structure). Besides the aforementioned substitutions, there are also included rearrangements, insertions and deletions along the DNA strands. Rearrangements in gene sequencing are also important, since they may also change gene interactions to some extent. Insertions could be produced when a terminator base is interleaved, then ending the copying process (see Fig. 2 and chapter 1, as a reference). Deletion is the loss of genes that could occur either in chimps or in humans' DNA.

New genes in either species, is another important event that could be determined. This knowledge, in particular, may provide some light in the always-captivating contrast between chimpanzees and humans' physiology and pathology issue, simply explainable by a sequence difference. Furthermore, in such a comparison, the action of natural selection, which could occur in early human evolution, could lead to significantly changed genes in human's genetic endowment, compared to their counter parts in chimps' genome. Detecting those genes, according to scholars' criteria, is of the utmost importance since they confer the human unique-health conditions.

Other important research profiles are also envisioned in the project, including among others, the selection that occurred during the period after the divergence of humans from chimpanzees, probably comprising the living period of our common ancestor. It is understood that the most common allele in a population is the ancestral allele. If it can be proved wrong for a specific allele, or a certain region of DNA, then chances are that such allele, or region, is a resultant allele in human population by latest selection, which will provide a peerless knowledge on our most recent genetic changes, throughout our own evolutionary process. In addition, chimpanzee sequence will provide a tremendous tool to fill gaps in the human's sequence itself, to better understand the changes that have occurred in the evolutionary progression of our species, parallel to chimps' evolution.

First, when we talk about chimpanzees, as you already know, it is mandatory to specify, which of the two species the issue is about. Yes, remember that this matter was already discussed somewhere before in this book: the common chimpanzee (*Pan troglodytes*), and the pygmy chimpanzee or bonobo (*Pan paniscus*). The Chimpanzee Genome Project is being accomplished and progressing only on the common chimpanzee. Nevertheless, by 2007 the Max Plank Institute for Evolutionary Anthropology in Leipzig, Germany, was ready to start the bonobo genome. To my understanding, Bonobo's genome is probably more important, from an evolutionary point of view, than the common chimp's. When both accurate studies will be ready, science shall be in good shape to make unprecedented comparisons among common chimps, bonobos, Neanderthals and extant humans' genomes. A map-based assembly of the bonobo genome is on its way. Once the assembly is done the researchers plan to study the evolutionary history of both chimps and bonobos.

At first glance chimps and bonobos may seem very similar but, in fact, they are two different species, which according to mtDNA studies conducted, both species split some 1,500,000 years ago, actually in early Pleistocene; the bonobo's genome is about 0.4 % divergent from the chimp's. It was not until 1928 that the bonobo was confirmed as a different species, and in 1933 it was acknowledged the zoological classification as *Pan paniscus*, published in the American Journal of Physical Anthropology.

The importance of chimps and bonobos concerning human lineage lies on the fact that they are the closest living species. The comparison between both species of chimps and humans, may run regarding different aspects: behavior, anatomy, physiology, and evolution. Chimps (both species) during their roughly eight million years of evolution, since they split off from our common ancestor, have not changed drastically, and a rather stable evolutionary condition has been observed, instead. It is startling to admit such a huge lapse without any progression, at least of any importance in evolutionary terms. Besides, no fossils have been found to grab the slightest comparison, as evolutionary biology demands, of the evolving phenomenon in chimps. Such an absurd situation precludes any possible consideration that may serve as a basic approach to understand chimpanzee's evolutionary history in particular, and regarding all living apes in general. These circumstances also preclude any reasonable approach to the knowledge of the common ancestor's appearance.

At this point, it is necessary to call the attention of the reader on the geological period between 22 and 6 million years ago, known as the Miocene, which was characterized for the huge proliferation of apes, that is why paleoanthropologists consider such an epoch consistent with a real "planet of the apes." It is believed that at the time, more than one hundred ape species thrived on our planet. From those ancient apes, the human lineage started up. Nowadays, only a meager number of apes survive. The great apes, gorilla, bonobo, chimpanzee and orangutan are well known to general public as they comprise one important show in major zoo parks around the world. It is opportune to say that humans are more closely linked to the gorilla, chimpanzee, and bonobo than they are to orangutan. For any layperson, it is evident that orangutan's external features set them apart from the rest of the great apes. Orangutan is red-haired, rounded face, due to fat pads encircling its face, moves around slowly, never showing the agility chimps, bonobos and gorillas display. They show a clumsy knuckle-walking pace, very typical of the species.

Behavior

The common chimp's behavior is by no means that of a peaceful creature, on the contrary, intra-group behavior as well as inter-group is rather elitist and aggressive. Ultimately, manifestation of extreme aggressive behavior has been observed in chimps to unprecedented levels.

Usually, a group of males patrol the borders of their territories, whether to prevent intruders or to observe neighbor groups' territories, which potentially could be invaded, and eventually be killed. It has been film documented the slaughter of a young female by a group of invading males from neighboring areas. After completely dismembered, the chunks of the carcass were shared by the invaders, in a bloody feast manner. For anthropologists, it is not clear the purpose of such a behavior. It has been invoked the extra supply of high quality protein to be included in their diet, but this explanation seems to be too simple to account for a whole strategy of spying and then assault, which is somehow risky, just to obtain proteins. If it is the true motive, then we have to admit that cannibalism is a customary practice among common chimpanzees.

It is surprising the results from studies where chimps are really devoted to hunting small mammals. This had been thoroughly documented since 1963. They usually add meat to their fruit and vegetables diet. This point is relevant from an anthropological point of view, since this meat preferences and inclusion in chimps' diet demonstrate that it was not *Homo* species the ones that for the first time supplemented their diets with meat, it was probably long before the split from the common ancestor that this particular behavior was established.

Nevertheless, the hunting behavior among chimp populations is not the same. In some groups, chimps hunt independently, in other groups hunting parties are organized, being some individuals strategically placed to lead the prey toward the ambushed males; other males prevent the game from escaping. As we can see, a real and smart strategy is entirely developed by certain chimp populations.

It has been well established that young female chimpanzees leave their group to join neighbor groups. This behavior prevents dangerous inbreeding. Nevertheless, if those females are rearing offspring, males of the reception group will kill the juveniles. In this way, the introduction of new genes is completely blocked, since newcomer juveniles will never reach maturity. Besides, this *modo operandi* simply prevents a prolonged anestrous female. The female will be shortly in a new estrus period, ready to assimilate genes from the adopting population. On the contrary, all males remain in their birth group; hence, a strong bond is easily developed to keep them together as a faction, in the style of a society, where strong alliances are built to endure. It becomes that all males in the same group are related.

What follows is a comparative approach to contrast behavioral aspects of Chimps and Bonobos. This is not a comprehensive list of behavioral patterns.

Chimps. Occasional bipedal walk. Are rude and quick-temper. Males show strong bonding. Often raise their hair. Use physical violence. Aggressive behavior. Hostility is common among groups. Warfare, cannibalism, and infanticide. Grand scale confrontations are common. Not consistently gregarious as a species. Females receptive only during estrus. Males control the groups. Poor vocal performers. Sex life is dull and simple.

Bonobos. Bipedal walking more frequently seen. Receptive, sparkling, and jumpy. Females show strong bonding. Rarely raise their hair. Seldom use physical violence. Aggressive behavior is rather mild and not elaborated. Hostility is uncommon among groups; they usually get along. Warfare, cannibalism, and infanticide have not been observed. Not grand scale confrontations; when fighting it is one-on-one basis. Gregarious; several groups may get together at night. Females sexually active almost during the entire cycle. Females control the groups. Good vocal performers. Many ways and modes of engaging in sex.

It is evident that such behavioral differences set them apart in their social environment. As a general profile, bonobos are more peaceful, easy going, and extremely attached to the group. They get along with each other in their society, and sex is not only a reproductive issue, it is used to calm the stress and create social bonds. Besides, as female bonobos are sexually receptive for a long period, demonstrated by their genital swelling, and they copulate with different males in the group mainly to appease and trade sex for food, there is no reason to practice infanticide by adult males, as common chimps do; since female sex is shared with different males, hence, paternity is not granted at all.

On the other hand, it has been demonstrated that female chimps, strategically, use copulation calls looking for support and protection from males against other aggressive females in the group. These copulation calls increase in frequency when high-ranking males are nearby, and chimp females hide their sexual activities when other high-ranking females are around, since competition among female chimps are actually dangerous. It is understood that this strategy sports by female chimps may lead to secure offspring from high-ranking males, which translate into strong and healthiest fathers.

These are the conclusions scholars arrived to, after careful studies in the wild and in captivity as well. Sex in bonobos is a *modo vivendi,* and not only oriented to reproduction, which confer a very different social structure to their groups and communities, very different from the common chimpanzees'. Besides, they are gentle and intelligent, chimps are noisy and quarrelsome. One aspect that ought to be stressed when contrasting bonobos and chimps' behavior is the fact that a ranking order is not important to bonobos; it does not really exist as such. Bonobos lack the order of dominance and submission; the dominant status is not an important feature in bonobos' society as it is in chimps'.

As females are important in bonobos society, it is suitable to confer certain degree of significance to selected females. First, if there are very important females (vif) in bonobo groups, they are respected not because of physical intimidation but seniority; it has been established that the older the female the higher the status. This trend is comparable to similar account in tribes and primitive groups in modern humans; the older is respected and considered. On the other hand, low status females are generally new immigrants from neighboring groups. These females' behavior may be considered as poorly conspicuous, they try to be not showy or noticeable within the group. Usually, these new-comer females create a quick bond with one single resident fe-male, to whom they offer special attention and care. It is said that if the resident female reciprocates, then a friendship status is established.

We can understand the importance of females in bonobo societies from the point that males' alliances are very weak, or less developed, hence females may apply a greater influence. Fol-lowing this, any young adult male could reach a top position if his mother is highly considered within the group. Conversely, those males whose mothers have died are usually put in the low-est consideration. Usually, in bonobos, mother-son bonds are long lasting. When arriving to a feeding site, males usually get there first not because they are dominant, it is just to get some ad-vantage over females; when females arrive males have to disperse themselves for the females feed as they please. Even low ranking females may displace the males on a feeding site. Infants are al-ways allowed to share the food. Males have to content themselves till the females and infants are satisfied with the meal.

It is an evident matriarchal society, where males are of a second importance. This might be the reason why aggressiveness is not common in bonobo groups. Females know how to handle social problems preventing any eruption of violence within the group. As has been proved through several researches, females are cunning and persuasive, using sex as a modulator to appease any quarrel that could be deleterious to the cohesion of the group, and of course, harmful to families' structure in their so-ciety. We have many examples in human history where outstand-ing women have been the utmost leaders to solve very complicat-ed problems, concerning politics and socio-economical values.

On the other hand, chimpanzee's share of food is very dif-ferent. If a male is the possessor of the food, first he will share with other males probably to create political bonds aimed to get cooperation in-group hunting. Bonobos do not set up hunting groups; hence they do not need this kind of sharing. Male chimps will also share food with certain females, probably intended to gain sex favors. Besides, those females receiving a share of food will hand it to their offspring, which will benefit from the male sharing attitude. Those infants could be descendants of that same male. Oppose to it, as male bonobos do not rule food control and sharing, females do not have to wait for a portion handed from the possessor male. Females control food sharing and allow infants, on an equal basis, to access it. Nevertheless, it has also been observed that if a male is in possession of food an estrous female may approach to have sex and then, be rewarded with a share of the food. As bonobos females begin estrous just one year after parturition, they are ready for reproduction sooner than female chimps, whose estrous cycle is resume around three and a half years after parturition. Besides, bonobos females may have sex albeit they are not in an estrous cycle.

It has been demonstrated, based on experiments in captivity, that bonobos can adopt the viewpoint of someone else, it is, they can put themselves in someone else shoes, and get ready to help other members of the group, in case of difficult situations. Besides, Kanzi, an extraordinary studied bonobo, learned to use, in a correct manner, the American Sign Language, by which he could request a multiplicity of things, including, of course, food, play, toys, etc. Furthermore, he was provided with earphones, and he was asked, from a room next to his, to pick up from a bunch of pictures a particular one, and he did it perfectly, no mistakes, no hesitation. This experiment demonstrated that Kanzi was able to understand our language after a short training. He could memorize the pictures according to our language. Beyond that, he could create simple sentences to request or express basic needs or desires. Remember that our babies learn the same way; we point to certain objects and name them, just to create an audio-visual context for the baby to learn and identify such objects. Of course, not all bonobos behave as Kanzi did, but such individuals as Kanzi are precisely the *avant-garde* within a population, and their offspring may inherit such characteristics.

I would like to introduce an extraordinary example of a common chimpanzee, known for the public by the name of Oliver and also as "Humanzee," a lexical combination of human and chimpanzee. First, he walked completely upright, physically he was different from a common chimp, and his head was small compare to any chimp, his manners by no means match with those of a chimp's, his eyes color was different, he was not aggressive as male chimps use to be, his ears were projected backward and finally, he did not get along with other chimps, he rather preferred humans,. In fact, albeit Oliver features are uncommon, researchers did not effectively consider how useful he could be to explain certain aspects of evolutionary importance, regarding apes and humans. Furthermore, as a rule, research institutions have ignored this rare ape. There are no records of x-rays studies from pelvis and femur bones to give an accurate answer why he can walk upright with no apparent difficulty. Could Oliver, if seriously studied, shed some light on the common ancestor we shared with chimps? It is a very uncanny contrast if we compare all the efforts paleoanthropologists make to search for fossils, and this living ape has been sent to the oblivion in scientific terms. Furthermore, the Yerkes Primate Center in Atlanta, an outstanding primate research center, through a qualified spokesperson, made publicly the assertion that specialists in that center never heard about Oliver.

If science is always longing for a clue to elucidate any aspect of apes and humans' evolution, why was Oliver discarded as a creature not important enough to provide significant information on evolutionary grounds? In addition, a report based on Oliver's DNA concluded that though it looked like a normal chimp's DNA, something was different in the sequencing from *Pan troglodytes* species. Unfortunately, the report was not accurate enough to pinpoint what the difference was.

Probably, I dare to speculate that not all hominids were able to start the difficult process of language onset though all biological conditions were ready to promote the first steps toward the creation of a language structure. Consequently, natural selection favored the most advanced ones in language production, since at a certain specific stage of human evolution, language was necessary. If we take for granted that chimps and bonobos have stayed with no outstanding evolutionary changes since *Homo* diverged from them, some six million years ago, then we may assume that our common ancestor probably had the same capacity, or very similar, as bonobos have. This means, the primeval basic of language was probably already present in that unknown common ancestor. I do not infer they were able to talk, but the basic of communication beyond the grunt, was probably present in those unknown hominids; they were lost forever millions of years ago, in the inscrutable darkness of time.

We developed such capacity up to the point of a complete language structure, with full communication possibilities; chimps and bonobos only withhold that basic capacity, but did not go further, why? Possibly, the answer will never be known, since the full explanation to understand the acquisition of language comprises a complex amount of events, including biological needs, environmental realities, and mutations that led to additional steps in accomplishing the phenomenon of language itself. This fact put us apart from *Pan paniscus* and *Pan troglodytes*, and eventually touched our species with the most outstanding element in any living creature, the full thinking-language process. Thinking, though rudimentary, was the first acquisition, since the process of thinking ought to precede the acquisition of language; there cannot be language if thinking is absent.

Some specialists believe that the pressure on the evolution of complex thinking, finally found a way out in the outcome of language, it was like an outburst strongly pushed by the immense necessity of expression, communication, and understanding, in the evolving human species. One of the controversial points is to explain why the chimps and bonobos did not undergo the pressure of evolutionary thinking, albeit we share a common ancestor. The "out of the forest" hypothesis is not so convincing, as has been explained somewhere before. Besides, when *Homo* split from the lineage of chimps and bonobos, those hominid forms were basically similar to the genus *Pan*. The most mysterious event is exactly, why at present we share more than 98% of our DNA after six million years, and we differ so drastically? If the divergence is the amount of sharing at present, then we were 100% similar DNA at the moment of divergence. Less than 2 % has been more than enough to explain the difference between a "beast" and a human. If two DNA samples, one from a human and one from a chimp, are given to a researcher without any evidence of the source, and then the researcher is asked to determine the genetic relation between the samples of both species, probably the verdict shall be that the samples belong to extremely closed species, probably very similar in appearance. What a *fiasco* for science foundations! Fortunately, for scientists, we are extant humans and chimps are extant apes.

It is difficult to conceive that with less than 2% difference in DNA the ability of rational thinking and language, besides other morphological and not less important features, set completely apart humans and apes. Nonetheless, there are similarities in our common behavior. Chimps never approach other groups to develop friendly and peaceful relationships. This behavior seems to be inherited; it is part of the behavioral endowment in chimps. If we recall ancient groups of humans as the Barbarians, among others, they behaved just as chimps do. The focal point of their social idiosyncrasy was to subdue, fight other groups, and get advantage, very similar to chimps' well-known behavior.

Barbarians were not friendly with other groups; they were not the exceptions in ancient history. On the other hand, friendly aspects of bonobos' behavior are quite similar to our own, including sexual behavior. I dare to speculate that the common ancestor probably displayed a behavioral pattern, which included chimps and bonobos' behavior, which in turn was shared by then newborn *Homo* lineage. Nonetheless, it is not rational to ascertain that every single detail on behavioral grounds was already present in the common ancestor.

During adaptation, bonobos, chimps, and humans developed their own behavioral characteristics *de novo*, not present in their ancestor, though it is not far from the truth to consider the common ancestor's behavior a genuine blending of chimps, bonobos, and humans' behavioral patterns, to some extent. In anyway, speculations are the only possible tools to approach the **hard to explain** evolution of the ape and human lineages, so similar in genetic terms and so different in rational grounds. It would be just like the title of a science fiction movie if we come up to the topic as saying "The Man and the Beast Kinship." Precisely, it is the core of the issue, though it seems completely unexplainable. Are we similar beasts with different attributes? Are we another version of a beast just because we are "rational"? Is rational thinking and language, as a by-product, ground enough to set us apart from the concept of anthropoids? Do we call them anthropoids just because we believe we are the center of the "evolutionary process," which reinforce the anthropocentric idea on Earth?

Several scholars consider the bonobo a living model to explain many aspects of our own behavioral pattern; hence, it is considered an anthropological living gift that should be thoroughly studied, for them to provide us with peerless information on important aspects of our own evolution as a species; as stated by specialists, the bonobo research has just begun.

I would like to ask those scientists that are immersed in reclassifying zoological species, based only on molecular genetics results, if they would find a couple of lizard species with less than 2% genetic differences, what the conclusion would be? Are they sub-species? Would those scientists set them apart in two different taxa? I mention this, for nowadays there is strong, and in many ways conclusive tendency, to classify zoological species based exclusively on molecular genetics results, sequencing, regardless of ethological and ecological characteristics. In my opinion, this is a new expression of the old *reductionism*, a ridiculous simplification of biological phenomena, so severely criticized by advanced biologists, three decades ago. The point here is that from an ecological point of view, chimps and humans completely differ. And, our behavior, though somehow similar to apes' in some aspects, cannot be classified as alike. The contrast is too strong in both considerations. But when the issue is approaching other zoological species, the contrast among ethological and ecological characteristics, might not be so evident. To exclude both ethology and ecology from the classification seems to be absurd, simplistic, and even ridiculous.

Furthermore, the common chimpanzee and the bonobo's DNA might be identical, in general terms, but they are different species; this is a scientific fact accepted all over the world. In the case of these two apes species, DNA results fall short to determine the real taxonomic status. Other aspects as their social structures, habitats, and mainly behavioral patterns, set them apart in two different taxa, regardless of the striking similarity in DNA structure. Is not this example enough to discredit the exclusively sequence-based analysis to seriously determine a taxon status?

Biology is a complex science, many aspects are involved in any manifestation of living creatures, and when we are dealing with multifaceted evolutionary pathways we have to be very careful not to take the easy trail to have quick results, booming figures, and quicker publications to secure future grants. These, at the end, will lack many essential aspects, which are the genuine key to completely understand the subject under research.

Additionally, we have to be very careful how we treat our closest relatives, chimps and bonobos. If we are less than 2% different in our DNA, is it a humane treatment to have them in zoo parks just because they cannot speak and express their feelings as we do? Would you imagine having your brother or your cousin as a show-specimen in your backyard with all the commodities and food, and people paying to see them? That is exactly what we are doing with chimps and bonobos. I was very specific in previous chapters when considering the so-called common ancestor. Then again, if we have a common ancestor shared with chimps and bonobos they are our brothers and sisters, not merely our closest relatives, as the most conservative scholars were prompted to name them. As we consider *Homo sapiens* the *non-plus-ultra* of evolution, or creation, then we adjudicate ourselves the right to cage chimps and bonobos for zoo visitors to watch them, and laugh at them, just because those speechless apes look funny. Unfortunately, they cannot tell us if they do agree to be "a living show" for zoo visitors, or if we look, from their judgment, degraded, weak, and abnormal almost-naked apes, with protruding lips, and fuzzy eyebrows.

Besides, no matter how much space and terrain those apes may have to play and wander about in any zoo, they will never develop the true pattern of their behavior as in the wild. If they are caged, which is common in many zoo parks, stereotype behaviors are shown. Usually, boredom is the most common reason for such stereotyped behaviors, but lack of social contact with members of its own species is also a common cause that triggers such patterns. Gorillas are frequently kept in individual cages. It is very familiar to zoo visitors observe the repetitive pacing of caged apes, going to-and-fro, which is the only means the restricted animal, has to exteriorized its uncomfortable situation. Furthermore, the innate exploratory behavior in chimps and bonobos is completely cut-off, due to the utterly restricted enclosure they live in. As has been masterly stated by scholars, it is fascinating to observe the striking similarity between young chimps at play compare to children. Both species, while young, take any toy or similar device to eagerly thump it against the floor, pull it apart, throw it away and then hand it again to continue turning it to pieces.

All these behavioral patterns are expected in chimps during their first years of life. Unfortunately, zoo enclosures deny them the normal behavioral development the species requires; it is not far from depriving children of the ordinary joyful play they deserve to develop typical behavioral patterns when adults. In few words, it is not reasonable to prevent the normal behavioral development that our ape relatives deserve.

They need an accurate re-creation of the jungle, which is completely impossible for a zoo; the visitors cannot observe them otherwise. That is exactly what makes miserable the lives of our cousins or brothers, whatever you prefer to call them. If there are some trees in the space provided for them, usually the trunks of those trees are wrapped in some metal sheath to prevent the apes to climb freely, being this activity the most remarkable behavior that characterized apes and monkeys as well.

Take a tour to different zoo parks in big cities around the world, or just in the United States, you will see by yourself what a disgrace is the enclosure and captivity for apes. Hello! At this point molecular genetics are not being taken into account; we show them in zoos as beasts not as our poor relatives deprived of the speaking gift, that natural selection...evolution...creation, gifted us with as the most outstanding privilege on Earth.

This is a very clear example how *Homo sapiens,* in general terms, subdue, enslaved and abuse the weaker; no matter if they are our closest relatives, as some scientists prefer to call them.

Besides, it is well known bonobos and chimps have been slaughter in Africa as a source of food, by the tribes that share the same ecological environments. Fortunately, many efforts have been accomplished to protect them in their natural habitats, and some achievements are the result of such efforts. Nevertheless, many specialists consider the bonobo as an endangered species. Probably the only fact that helps the existence of the bonobo populations is their location in Zaire territory. The bonobos' distributional area is very difficult to access. Lomako and Wamba are two important research sites, and in particular Wamba is a very difficult region to reach. Nonetheless, bonobos have to be protected, and if possible, reproduction in captivity might be mandatory to perform an increment in bonobos' world population. Those individuals born in captivity must be released back in the jungle, where they belong, using appropriate methods and procedures.

Specialists have openly declared that just a moderate success, regarding apes management, in general terms, has been accomplished during the twentieth century. The first available records of captive apes to reproduce were chimps in Rosalia Abreu's colony in Havana, during 1915, almost a century ago. She was a lover of apes, and she was wealthy enough to keep them in splendid conditions. Coincidentally, an outstanding bonobo named Prince Chim died of pneumonia during a visit to Madam Abreu's farm in 1924. This exceptional bonobo was carefully studied and observed. He was very intelligent, all the time imitating the actions and gestures of his human friends; a well level-tempered individual, inquisitive and friendly. Unfortunately, at the time, ethology was not an established science, only sporadic reports on certain species, sometimes poor in details, were available in very scarce cases.

Regrettably, many laypeople are reluctant to accept that those hairy creatures are not very different from us. In many ways we share similar feelings, as depression, enjoyment, sadness, and rage, just to mention a few; all this is well known and understood by ethologists and keepers in zoo parks as well. Common people, without a Nature culture, are prone to believe that animals are just animals, and human beings are definitely superior, no question about it. It is one of the most painful dilemmas regarding the integration of humans into Nature. We came from Nature, we are what we are because Mother Nature was, is, and shall be the cradle of life. Without the prolific Nature our beautiful planet would be nothing but a sterile deserted insignificant mass in the immense universe. If natural selection and then evolution were responsible to settle down life on earth, Nature was the cradle that pampered life. If creation is the reason for life, then again, Nature was the cradle chosen by creation.

The Divergence

At the time of divergence, a hominid lineage sailed off from apes, as an incipient *proto-homo* gene pool not well understood by science today, since we have no idea how those archaic proto-humans could be. Probably they enjoyed a well diversified genome, plenty of genetic possibilities and sporting the ancestral well established genetic endowment plus new gene recombinations and mutations that made possible the leap forward, from apes to a very primitive form of a proto-homo. That was the very beginning of our existence on Earth. As mentioned before in previous pages, during the Miocene a plethora of apes roamed the planet. One of those species, perchance more than one, evolved toward the future genus *Homo*, who finally was divided in two main populations: those who left Africa, and those that remained there.

There is a discovery that is worthwhile to mention since the find itself changes, somehow drastically, our concepts and hypotheses of human evolution. In East Africa, not far from the slopes of the Moroto volcano, a single ancient lumbar vertebra was unearthed. The amazing point of the discovery is that the single vertebra there found, is remarkably similar to a human lumbar vertebra. Scholars cautiously assert that probably there is a misconception of what the vertebral structures of apes in the African Miocene era were like. This particular vertebra structure gave name to the apes that bear such anatomical configuration, *Morotopithecus bishopi*, some 21 million years ago. Probably, at that time in Equatorial Africa, as a main possible region, some mutation showed up, according to some specialists, that gave way to a completely new kind of body arrangement leading to the appearance of an upright hominid. This was just the beginning for the future millions of years of ape-hominid forms to come, until some 6 million years ago the split of humans and chimps occurred.

Now, the question is not exactly the consequences of such fortunate mutation, the real question, in terms of biological interest is, how many other mutations occurred on the way to completely create a finely finished human body? Then again, penguins are upright walking animals, but that is it. They are birds, but do not fly, they rather swim. What I try to say is that upright walking in itself is not enough to impulse all the complexities necessary to become humans. There are thousands of biological details, mutations, gene recombination, pleiotropic effects, and chromosomal rearrangements that should occur in a very delicate and specific sequence to really fulfill the requirements, to end up the task of the human body structure.

Another logical and almost mandatory question is, was *Moropithecus bishopi* the only ape species in the Miocene that sported upright walking? It is difficult to answer, but probably not, in the same way it is probable more than one ancient ape species gave way to the most primitive proto-homo. Remember that, according to scholars, Neanderthals and extant humans do not share genes. Then, we have to understand, according to those scholars, that we both *Homo sapiens* and Neanderthals, descent from different lineages. Then, the possibility of more than one ancient Miocene ape sporting upright walking is really open, besides the possible contribution of more than one of those species to the *Homo* lineage. No one can discard the possibility of two close ape species interbreeding to create a very special viable hybrid from which a rich DNA was the starting point to develop a whole lineage, which led the appearance of *Homo* after millions of years under evolutionary pressure. Precisely, a rich DNA is a fertile ground for natural selection to choose the optimal solution based on environmental conditions at the time.

As has been profusely discussed in former pages, only small groups of *Homo* emigrants took the trail out of Africa, in so small quantities that their gene pool carried the perilous messages of extinction, which did not occur. Nevertheless, so small populations possessed exiguous gene variations, which inhibited natural selection to act upon enough options to select the best out of the good ones. On the contrary, those populations that remained in Africa were large enough to guarantee a vast and rich gene pool, as they have nowadays.

Today, through different results from very accurate and serious research procedures, it has been demonstrated that African populations are more genetically diverse than Europeans are. This means that their gene pool is richer; they are adapted to their environment better than Europeans are to theirs. From a population genetics point of view, they have an edge in survival terms compare to Europeans. It is true, you do not have to be a scientist to realize that they are healthier than people in the so-called developed countries are, albeit many populations in Africa suffer famine, but that is not genetic, that is a social problem. They are exposed to many hazardous living conditions; they not only survive but also thrive.

This is not the whole story. Comparisons have been made to determine the variations within species, apes and humans. The results clearly pointed that two chimpanzees are more different, from a genetic point of view, than two humans are. Consequently, genetic variation is so much ample in chimps than it is in humans. We are restricted in our genome, they are genetically rich, and we are poor. Comparatively, apes suffer less health problems than humans do, just because our genetic-based diseases listing are enormous compare to chimps or bonobos. Through the six or seven million years of evolution that *Homo* and *Pan* walked side by side, after they diverged, chimps probably enhanced their gene pool; we depleted ours, just because our ancestors are nothing but a bunch of emigrants that left Africa. Chimpanzees and bonobos stayed on their proper environments, well adapted, thriving, and renouncing, with the aid of natural selection, to acquire rational thinking and upright walking.

Now, some confusing arguments may arise. If the emigrant populations of archaic humans left Africa, and of course, their evolution was accomplished far from their homeland (you may think about *Homo erectus, Homo heidelbergensis, Homo ergaster* or even *Homo neanderthalensis*), why so many well adapted species as these were, could not make it, and the last migrating group could populate the planet becoming *Homo sapiens*? (Cro-Magnon people according to fossils named). Furthermore, is it just because apes stayed where they belong that no obvious evolutionary pace has been accomplished for them, so far? And all these, in such a way, that specialists almost openly declare that probably the common ancestor was very much like the chimps, or quite so.

What kind of genetic endowment did the proto-*Homo* species have for an evolutionary progress to take place so manifested in them? Many hypotheses have been launched on the stage of reasoning evolution, just to explain the rather swift evolutionary trend that led to extant humans. In fact, no plausible ones have been elaborated yet, not to the point that could convince the most demanding academicians on evolutionary grounds. Once again, speculations in front of the complete absence of tangible proofs, or at least, basic proofs that could lead to a solid hypothesis, where no "illogical gaps" may stand as detrimental evolutionary profiles, which only shed doubts on the peerless path of our controversial species, all over the world.

Was it struggling for life that made brutes deserve an incomparable opportunity to evolve and reach the status that no other animal ever reached? Was it the closeness to extinction that pushed *Homo* forward as was never done before on any other species in the world?

But this is not enough, our cheap DNA compares to ape's, is another drawback in evolutionary terms. Apes were not, and are not in such disadvantageous genetic situation as we are, and probably we were that way since unknown ancient times. The "out of Africa" hypothesis has been discussed *in extenso* in this book, hence the reader is readily familiar with the implications this hypothesis compels on evolutionary models. Providentially, apes never went out of Africa, they would not be so successful, otherwise. If most of them are now a day endangered species, it is not because of natural selection acting upon; it is because extant humans are threatening apes' populations, slaughtering them for food, besides illegal trading. Then again, *Homo sapiens* acts as the most dreadful species on earth. Slaughtering chimps is not far from slaughtering human beings.

I may say that the true importance of bonobos in connection to evolutionary aspects has only been recently understood. Amid many debates and attempts to officially start, once and for all, a serious research on bonobos, one important aspect that unfortunately triggered back a conscious approach in understanding the significance of this ape, was the lack of adequate technology. It was not available in the last decades though plenty enough at present. But even so, a paper was published in 1979 where a careful body proportions study was established between Lucy (*Australopithecus afarensis*) and the bonobo. First, they are strikingly similar in overall body size, lower limb length, and cranial capacity. According to these features it appears that bonobos could provide a reasonable morphological connection between early hominids and the common ancestor, much better than any other living ape.

Furthermore, if proper researches are conducted, it is not surprising if a certain degree of clarification could be at hand regarding the possible events in the morphological transformation from ape to man, via the common ancestor.

As has been stated by some scholars, behavioral aspects and morphological evidences make almost impossible to deny that bonobos seem to be extremely useful in the construction of the path from ape to human. Even so, when comparisons were made to establish affinities among the living apes and *Australopithecus*, it came out that albeit bonobos share some important similarities, the common chimp, the gorilla, and the orangutan also share some few of them. According to the specialists that conducted such studies, the so-called common ancestor was not exactly like any modern ape; hence the best reconstruction approach should be conducted from all available living forms and extinct hominids. Additionally, when specialists try to establish careful comparisons among the living apes and the well known fossils of extinct hominoids, one aspect, one important aspect is always out of the stage: we do not really know if another ape forms contributed, decisively, in the long humanization process, and from all those possible forms, only the extant apes are accessible for us now to establish precarious comparisons and research.

The find that represents *Ardipithecus ramidus* is, among other important concerns, the revelation that in the completely obscure past of our evolutionary process, an entirely unknown number of ape forms, which existed during the Miocene, could be in our direct evolving lineage. Chimps and bonobos are only a couple of species closely included in our evolutionary tree; others, unknown for us today, may be the clue of the unexplained, sudden in time, appearance of *Homo*.

Some specialists have stated that the aggressiveness in humans could be explained if considering a genetic basis behavior somehow shared with chimpanzees, and probably also present in our common ancestor. This assertion has been strongly criticized by others, since no genetic evidence has ever been demonstrated regarding the existence of a gene associated to "demonic" or "violent" behavior. If it is completely true that such a gene has not been spotted, yet it is by all means reasonable that more than one behavioral pattern, at least to some extent, may be shared with chimpanzees, since we are more than 98 % genetically similar to them. If this high percentage of similarity means nothing in behavioral terms, then it is useless to be concerned in studying chimps and humans' genomes, since one of the important aspects that could be highlighted in such researches, lies on the possibility of a better understanding of our own behavior as a species. Of course, the study of both genomes is of a great importance for the advancement on primatology, anthropology and paleoanthropology, as well as for the genetic science itself, which is in the center of this extraordinary effort to elucidate as many aspects as possible regarding human evolution.

What science is truly facing is the hard to explain close genetic similarity and the extraordinary differences between chimps and humans. That huge difference cannot account for less than 2% in DNA sequence. Of course, in modern times the concept of epigenetic inheritance is widely discussed and considered, since this term means phenotypic variations, which do not depend on DNA sequence variations. This and other important concepts in modern evolutionary grounds will be discussed in the next chapter.

There is something else that cannot be explained only on a genetic basis, and that is exactly the point that many scholars try to approach through their particular disciplines. Very careful hypotheses ought to be established and at the same time find the courses of action to prove them. One of those points is the swift evolution of our intellect, up to the point that we are the exclusive creative species on the planet. It is utterly evident how the human species has developed all branches of knowledge in so a short time, in terms only of a few thousand years. That is not enough time to develop such an immense evolutionary achievement as our brain and mind are. On the other hand, chimps behave as clumsy toddlers compare to us. This cannot even stand the most elemental comparison based only on less than 2% genetic difference.

It is completely misbalanced to bring forward the physical traits in chimps, as long pointed canines and body hair that we do not already posses, in a futile attempt to demonstrate or persuade the audience that chimps' aggressive behavior is not related, genetically of course, with the outstanding aggressiveness typical in humans. According to the opponents of the human-chimp's aggressive behavior issue, the primitive aggressive behavior was lost along the evolutionary path, as well as the long pointed canines and the abundant body hair. But let me tell you, many modern human males exhibit abundant body hair, probably not less than 50% of that present in chimps. That hairy human phenotype is common. It would be really interesting to conduct a study to determine what percent or hairiness is present in that particular human phenotype compare to chimps' equivalent. This is another trait that could be explained in terms of ancestral inheritance via the common ancestor. In fact, we may share several traits with chimps, some could be evident, and others somehow cryptical, only adequate and scrupulous studies may shed some light on these controversial aspects of apes and humans' evolution. Open minded researchers is what is needed, not "buts" and "ifs," which at the end lead to nowhere, just because those extremely conservative, or fundamentalist representatives of science are somehow reluctant to go further; heaven knows what the reason is. To deny any kind of possibility, no matter what the essence of such topic might be, is to go blind in the path of cognitive progress.

Of course there is no guarantee that things that chimps do may not be pertinent to understand what, we, humans do. But even so, if they are our closest living kin, it is not illogical to infer certain chimp-human behavioral relations in evolutionary terms. To those that proclaim these behavioral patterns as almost useless or irrelevant, since those elements have been lost along millions of years of human evolution, or worse, to point that these behavioral elements have not been a human attribute ever, an appropriate answer is: no matter how many millions of years have been coursed; our relation survived that huge lapse, and stay to be less than 98 %. Accepting this, consciously, is the only way to accept that any possible relation may exist between

chimps and humans, hence, any attempt to understand our behavioral complexities by establishing a contrast with chimps, is
completely legitimate from a scientific point of view. Chimps and
humans' evolution has been a parallel process, from a common
ancestor that gave birth to both *Pan* and *Homo*.

A very controversial topic that has endured, perhaps for
centuries, is the abstract concept of *natural man*. Nowadays,
some authors, in fact just one or two, strike against the study
of chimps arguing that is non-sense to consider any study on
chimps if aimed to approach the primordial or natural condition of humans, understood genus *Homo*. Of course, chimpanzees are not the prototype of natural man, but in fact they are
the living closest survived relatives of the theoretical natural
man, it is, the most primordial form, unfortunately unknown to
paleoanthropologists since no fossils have been unearthed, for
whom no real culture could be talked about. Culture is entirely
a concept for a certain degree of evolution in humans, apes have
no culture. Euphemisms are not welcome in scientific matters.

Evidently, there is a gap between chimps and the first most
primitive *Homo*, who had no culture *per se*; unfortunately, that
unknown species has not been found. Nevertheless, scientists
are urged to infer that such a species somehow roamed somewhere, in Africa obviously, and it was not sufficiently evolved to
allow us to consider a genuine culture for such a primordial human form. From a sociobiology point of view, culture is the accumulation of traditions, from one generation to the next, always
increasing in content and quality, evolving together with other
fields of human knowledge, and fitting in a social context, in
time and space. Inconceivably, some specialists, fortunately few,
consider cultural diversity the trademark of evolution, or the
true evolutionary trademark. It is just like placing the horse behind the buggy. Culture is the direct product of intellectual development. Our brains, our minds create culture. If there is not
rational thinking, how can culture appear? Again, culture is the
product of human's mind, just as inventions are also part of our

culture. Why apes do not have a true culture? Simply, they are not rational beings. Culture is the product of developing consciousness, exclusive gift only humans possess. We are conscious of our own existence, apes do not, hence they do not have a true and genuine culture, just behavioral patterns, mostly inherited, and the capacity to learn up to the limits an ape can do. Rational thinking is the hallmark of humans' evolution as a species.

It is accepted that one of the most primitive *modo vivendi* in human groups was the hunter-gatherers condition, probably dating more than 10,000 years ago. I already mentioned in chapter 4, the Hadza, which live around Lake Eyasi, in northern Tanzania, as a nomad hunter-gatherers group in our time. Well, if we compare chimps' behavior, taking all the details and facts that have been studied by specialists, we have to accept that chimps develop that same living style as hunter-gatherers extant humans do. Recently, in 2010, it has been published an extremely detailed research on Congo's Goualougo Triangle, where a very sophisticated method of capturing termites displayed by chimps has been utterly documented. This group of apes uses their own tricks to access the inner part of termite mounds, besides developing a complex strategy to puncture the termite mound in different spots, and then using different sticks and maneuvers to get and eat the insects. This is hunting, very simple and riskless but it is. Of course, they are extremely dexterous in gathering food. Without a doubt, we may say that the most primitive human groups and chimps share the same ability known as hunter-gatherer way of living. That is exactly what both species did and do for a living, Hadza people are a living example.

Now on top of that, some authors have marginalized the issue of aggressiveness in humans as an ancient inheritance from the common ancestor we share with chimps, since no genetic proof has been yet disclosed. Yes, it is absolutely factual, the aggressive gene or genes have not been isolated as a material proof of such assertion, as I discussed somewhere in previous paragraphs, as well as the gene for long canines in apes has not been either isolated. But even so, if humans and common chimps are both aggressive, it is not parsimony to adjudicate different origins to such a behavior. We are very much alike, genetically. The simplest explanation may become the most plausible one. This is parsimony, a principle used in science.

As a mere and simple illustration of a "remote possibility" of certain predictions to come true from a certain time in the past, in human culture and technology, it is legitimate to mention that, if my grandfathers were told about the marvel of a cell phone at that time, they could answer for certain, based on strict orthodox science principle, that "no actual proof or evidence is available to indicate that such a device could be designed or constructed." This is exactly what many fundamentalist scientists' attitude is. They deny any possible approach or reasoning, just because there is no evidence of anything that could explain this or that. One thing is to be a serious and consistent scientist and another very different is to be a locked-minded individual. If you do not dare to open your mind to new possibilities, though they seem odd according to present knowledge of science, you will never make a step forward envisioning new unexplored horizons, just because you are an ill-conservative, fundamentalist scientist, scared by possible critics, or you are as blind as the medieval "scholars" that took for granted that Earth was flat; that was the orthodox knowledge that nobody dared to challenge at the time, just because of the Church or any other compelling beliefs stated the limits.

Christopher Columbus had plenty of courage to embark himself in the most revealing sailing adventure in human history, then proving the stupidity of a flat Earth and the reality of a land beyond. He did not believe that further the horizon line there was no land. Thanks to him we can talk about America. He just followed Toscanelli's ideas about the spherical shape of Earth. At the time, Toscanelli was not taken into consideration since he did not abide with the official accepted maritime concepts. After Columbus, many sailors dared to sail toward the mysterious far beyond. *Audentes fortuna juvat...*

What I try to introduce is the idea that, at present, the emergence of *Homo sapiens* is not well understood at all, hence, any attempt to disclose the most modest fact or evidence that could shed the dimmest light on the issue, is by all means encouraging though we may stumble a while in the "unknown" and quivering trail of daring assumptions. Perhaps one of them is the corner stone that many do not dare to touch, or even consider. More aspects of all these will be dealt with in the next chapter.

Some try to explain that it is aggressiveness, an element of chimps' behavior but never present in humans, or worse, present in the common ancestor but not inherited by humans. These "alternatives" according to a scholar, are crucial since the skill to distinguish among alternatives is testability in science. No, it is wrong from a lexical approach. Testability is the condition to be testable; the ability to distinguish among alternatives, is discerning not testability. This scholar strikes against the debate human-chimp associated behavioral patterns, throwing the essence of the discussion to the dominion of "metaphysics," not science. The open discussion of this feasible shared behavioral patterns, is inference, which is part of the scientific reasoning; statistical inference, for example. Finally and conclusively, some scholars stated, for the last time, that it is impossible to approach human nature from chimps, since they are not human. Of course, this is nothing new, they are not human as we are not chimps, but they are more than 98% humans, and we are more than 98% chimpanzees, genetically speaking, in spite of our cultural achievements.

Another aspect that I dare to consider is the "undisturbed ethological studies in chimps and bonobos." Yes, everyone knows that many efforts and researches have been accomplished focused on ethological approaches on both species. Nevertheless, the presence of humans, who have to be very close to chimps' groups to really observe behavioral patterns and interactions, may disturb the normal development of those expressions in free ranging and wild populations of chimps and bonobos.

The topic of the human presence as a disturbing component in the study of apes has been profusely discussed during years. What is the true limit these apes tolerate regarding the presence of humans without preventing a full expression of their innate behavior, or more intimate relationships, which could be a real clue to understand some of the human behavioral expressions? The answer, obviously is, nobody can tell. Probably, what we know from the years of persistent studies in Africa, are just the most common patterns, those that cannot be hidden since they represent the core of the species behavior, but this does not mean that other expressions of behavior might not be concealed in the presence of the human observer. How many of those patterns have been concealed? We do not know. But the question is still open.

They readily detect the presence of humans; this fact has been stated in several publications. Hence, as soon as they became aware of such a disturbance, many things could change regarding the way they behave, and prevent them to openly show all the details that really comprises the rich ethological endowment chimps and bonobos, our closest kinfolk, possess. They are apes but not stupid, probably we look very much to them as another group of weird apes; naked, tall, with protruding lips, carrying things (the equipment for camps and research), and walking upright with uncommon movements. They are cunning, not dim-witted jungle creatures.

In all about, chimps, bonobos, and humans may represent an exceptional "three-dimensional" product of a hard to understand result in terms of evolution. Moreover, though chimps and bonobos are not beyond the category of highly developed apes, so successful beings they are that they endured along the difficult path, side by side with humans since we split six million years ago; gorillas and orangutans as well, albeit these two species have not so a close genetic relation to humans as chimps and bonobos have.

The biological deterrent that prevented genus *Pan* to evolve, then going through a persistent unmovable evolutionary condition, is not clear at all. All the hypotheses carefully crafted trying to explain such a circumstance, have not been proved whatsoever. Furthermore, if we invoke the cunning condition of these apes, it is difficult to accept that no primordial elements were present during the last six million years to whim both, or at least one of the *Pan* species forward, into the path to higher forms of apes, almost at the borderline to humans. The jungle *versus* African lowland is not enough to explain the evolutionary immobility, since baboons are lowland monkeys and they have no evidence of upright walking, not even the least indication of such possibility.

Genus *Pan* continued an inalterable evolutionary passageway while many, we do not know how many, *Homo* species were thriving in different parts of the world. What does really represent that ill fated "less than two percent" DNA similarity between they and we?

Why that insignificant proportion stopped there without any further possibility of decreasing the difference to create them more similar to us? This question is quite similar to, why Neanderthals disappeared and we are here? They were already humans and unquestionably, well adapted. At this crossroad science crashes against an immense wall, that science itself together with its holding hands, the scientists, are impotent to slide an acceptable explanation. I say explanations for hypotheses are useless since they will never be proved; there is no way to do it.

Then again, many questions are on the stage, and many more will arise with every single unearthed fossil yet to come, simply for there are no possible expectations to elucidate every step in the unknowable evolutionary process from apes to humans on Earth. I do not want to sound as agnostic, but critical thinking indicates what must be expected, unless we, somehow, are devoted to scientific miracles. We should "reconstruct" certain processes based on suppositions; I may say smart guessing, the cheapest and more handy tool that we may use in paleoanthropology, archeology, and of course in genetics, and in all related biological fields that may contribute to elucidate no matter what simple a detail might be. The point is to be focused on every possible key, and of course, as I stated before, not to be too conservative just because hard criticism may hurt or may turn to be detrimental in the scientific realm.

Recommended Bibliography

Boesch, H., and Boesch, C. 1994. Hominization in the Rainforest: the chimpanzee's piece of the puzzle. Evol. Anthropol. (3) 1: 9-16

Coolidge, H. J. 1984. Historical Remarks Bearing on the Discovery of *Pan paniscus*. In The Pygmy Chimpanzee. Evolutionary Biology and Behavior. Plenum Press. N. Y. 435 pp.

De Waal, F., and Lanting, F. 1997. Bonobo. The Forgotten Ape. Univ. Calif. Press. 210 pp.

Ely, J. J., Leland, M., Martino, M., Swett, W., Moore, C. M. 1998. Technical Note: Chromosomal and mtDNA analysis of Oliver. Amer. J. of Physical Anthropology 105(3):395-403

Filler, A. G. 2007. The Upright Ape. A New Origin of the Species. New Page Books. A Div. of The Career Press, Inc. N. J. 288 pp.

Finkel, M., The Hadza. 2009. National Geographic. 216(6): 94-118

Foer, J., The Truth about Chimps. 2010. National Geographic. 217(2): 130-145

Gagneus, P., and Varki, A. 2001. Genetic Differences between Humans and Great Apes. Mol. Phylog. Evol. 18: 2-13.

Kaessman, H., Wiebe, V., and Paabo, S. 1999. Extensive Nuclear DNA Sequence Diversity Among Chimpanzees. Science 286: 1159-1162

Kano, T. 1982. The Social Group or Pygmy Chimpanzees (*Pan paniscus*) of Wamba. Primates 23(2): 171-188

Lang, E. M. 1971. Experience with Breeding Apes in Basel Zoo. Proceedings of the International Symposium on Breeding Non-Human Primates for Laboratory Use. Berne

Marks, J., What It Means to Be 98% Chimpanzee. Apes, People, and their Genes. 2003. Univ. of California Press. 312 pp.

Mc Henry, H. M., and Corrucini, R.S. 1981. *Pan paniscus* and Human Evolution. Amer. J. of Phys. Anthropol. 54: 355-367.

Quammen, D. 2010. Jane. Fifty Years at Gombe. National Geographic. 218 (4): 110-129

Shea, B. T. 1981. Comment on Bonobos: generalized hominid prototypes or specialized insular dwarfs? Current Anthropol. 22: 368-369.

McHenry, H.M.1984. The Common Ancestor. A study of the Postcranium of *Pan paniscus, Australopithecus,* and Other Hominoids p. 201-230. In: The Pygmy Chimpanzee. Evolutionary Biology and Behavior. Plenum Press, New York. Susman, R. L. (Ed.) 435pp.

Tutin, C., and McGinnis, P. 1981. Reproductive Biology of the Great Apes. Academic Press, N.Y. pp. 239-264

Zihlman, A. L. 1979. Pygmy Chimpanzee and Early Hominids. South. Afric. J. Sci. 75(4):165-168.

Chapter 5
Ecce Homo on Earth

The Approach

Our planet and we are undisputable elements of the universe. As such, we are, and our ancestral forms of living matter had been, subjected to four fundamental forces. Primarily, the calibration of nuclear weak force (responsible for radioactivity and the formation of basic elements in stars) and that of gravity are so accurate, that any infinitesimal difference, approximately one in billions, can misbalance the universe as we conceive it in its present route. Then, if any of these two forces would become just slightly weaker or stronger, catastrophic events will occur such as hydrogen conversion into helium, the destruction of hydrogen would lead to the absence of protons due to the decaying of neutrons, just in the early stages of the universe. An overall effect would be the absence of water or extremely short-living stars. Then, solar systems as ours, would linger a fraction of their cosmological time.

Electromagnetism, another important and exclusive force, stronger, much stronger than gravity and only acting on particles with electric charge, needs to be finely tuned for stars not to be too cold or too hot, making possible the appearance of life as has occurred in our planet. Finally, there is an essential ratio, electromagnetism/gravity, which is critical for planets and stars to be born. Besides, in the micro world, there are certain fundamental values that stay unchanged for the universe to exist: the mass of the electron is, exactly, half the difference between protons and neutrons' masses. This quantitative factor makes possible the stability of atoms that form the basis of chemistry and biochemistry, on which life is based. Besides, the strong nuclear force keeps tightly together protons and neutrons in the nucleus of any atom. This is the energy source of the sun.

Nevertheless, the components of the universe, in its incommensurable vastness, are not either eternal. Our sun is the center of our solar system, which is just one in millions in our galaxy, the Milky Way. Giant galaxies as ours can create and keep hold of essential elements as oxygen, iron, hydrogen, and many others, which are the building blocks of planets. Stars are not forever, when one of them explode in our big galaxy (very different is for smaller galaxies) all the elements resulting from such an explosion are somehow retained by interstellar dust and gases. They are hold back by the Milky Way huge gravitational field. In this way, they become elements to enrich the course of gas clouds, then initiating the process of new stars and planets formation.

At the center of our galaxy there is an enormous black hole, known by the name of Sagittarius A-star. Black holes, usually, are generated when a colossal star implodes, not before all its nuclear energy has been burned out. It is of a tremendous gravitational force, hence, it does gulp down errant planets, small particles, gases, nothing escapes from its huge and magnificent absorbing force, not even stars; they die in the black hole. Our sun stays twenty-seven thousand light-years away from it. Amazingly, this massive black hole can, and in fact it does, propel stars away. It is just like a cycle, the whole galaxy is in constant change, entering and re-entering different forms of cycles.

Nonetheless, the center of the galaxy is a true fruitful space, since stars get compactly together all around it. In this way, they create heavy elements necessary for life. Near our sun (in cosmic distance), a lot of newborn stars show orbiting disks made of gas and dust, which are the raw material in the formation of new planets. This event, precisely, is the *sine qua non* pre-requisite for the materialization of life forms. This is a continuous process, not easy for predictions to foresee the end of. Energy changes from one form to another, apparently, a never-ending process. Who can predict the end?

Our galaxy sustains at least one planet with intelligent life on it, our beloved planet Earth, where *Homo sapiens* came to be. According to calculations, our planet was formed around 4.6 billion years ago, from whirling solar interstellar clouds of dust and gas. The dust particles collide in a constant chaos, until the time comes to form larger chunks of solid matter, which in turn collide among themselves to form even larger bodies of matter, and collisions kept on during millions of years. Then, already huge masses of solid matter began orbiting the sun. This is just a rough and simplified scenario of the events that gave birth to planets, including, of course, ours. Gravitational forces played an important and unique role in all these happenings.

Recently, February 2011, the results of a cosmic census were discussed at the American Association for the Advancement of Science during the annual conference in Washington. The scientists informed, according to extrapolations based on the NASA Kepler telescope that not less than 50 billion planets exist in our galaxy. Besides, estimated numbers show a ratio one out of every two hundred stars (1/200) sustain planets in the presumably habitable zone, all over our galaxy. As scientists stated at the conference, this is just an approximate estimate of the ratio. At present, the accepted figure for the amount of stars in the Milky Way is very close to 300 billion. Of course, not all the stars have orbiting planets. More astonishing is the figure of the amount of galaxies in the universe, 100 billion galaxies.

It is necessary to have, at least, an estimate how universe is about to really understand the immense complexity and the extraordinary event that represents the emergence of humans on Earth. Astounding amounts of coordinated events took place during millions of years including physical, chemical, biophysical, biochemical, and biological, with all its branches; the final result is *Homo sapiens* on Earth. In fact, the extremely coordinated events, that brought life to earth and sustains the universe, are of a extremely fine-tuned sequence and reactions, as has been already given details of in previous lines. Some physicists try to create models to reconcile, somehow, scientific and belief systems regarding this astonishing avalanche of flawless sequential events.

The point is not to view certain evidences as religious facts; the core of the issue is to realize that those are fundamental energies, that we at present are not able to explain their nature. It is not a religious issue whatsoever; this phenomenon ought to be focused as another aspect of human existence. Such energies, albeit unknown to be clearly explained, are as real as our own thoughts, difficult to grasp their nature, but their reality is unquestionable. It seems obvious that a great amount of phenomena in the natural world are missing exploratory attempts to develop superior principles, apart from those of classic physics.

The new paradigm of a superb organized-creative universe is sustained based on cooperative properties of a new quality, capable of arising exclusively when both energy and matter reach a combination of higher complex states, not well understood at present, including forms of life, with the addition of consciousness. Besides, the concept or idea of a pre-conceived plan that the universe, itself, comprehends during its own development, is pointing to a teleological frame. It is, events determined by their ultimate purposes and not only by factors of causality. This extra-new model is adventurous, in the same measure that it shows a manifest propensity approaching to explain biological phenomena based on physics. Nevertheless, many evidences demonstrate that albeit great advances in essential science have been accomplished, numerous events in nature still require a different approach out of the accepted official establishment of physics and biological sciences.

If science has detected relevant cosmic principles of an incredible self-organization, unstoppable development, and the certainty of a designed universe, besides the evidence of a holistic structure, then humans could be a direct and unquestionable product of a whole cosmic or universe blueprint. One thing is obvious, probably hundreds of thousands of planets in our and other galaxies are fitted to sustain life. Until present we have no proof whatsoever, and in the future neither technology nor human existence will assure such proof; it is objectively speaking. I am not saying that among the 300 billions stars in our and other galaxies there might not be intelligent life. The question in front of such reality should be: why is life so disperse in the universe preventing any chance of contact? Is there any paradigm to explain why we are thousands or millions light-years away from our possible neighbors? Is it part of the universe's purpose blueprint?

Recently, fossil microbe forms have been found in small aerolites that crashed on earth. This kind of aerolites is rare, specialized institutions in USA preserve just few of their kind. Some specialists determined that these microbes are quite similar to those existing at present in our planet. This is just a primitive form of living organisms, but in no way it is ground enough to ascertain that life came from outer space. The only direct conclusion, according to this evidence, is that similar primitive forms do exist in other planets.

Science has struggled to create a single model capable to describe every particular aspect of the universe where we belong. Up to present time, it does not seem possible that such a model can be achieved. Modern times arrive with a network of related theories, known as M-theory. Every theory in the network is excellent, almost, to describe a particular phenomenon up to some acceptable extent. Let us put it clear, no single theory in the system can describe all aspects of the universe. The network of theories is the best approach. The M-theory let the possibility of different or multiple universes, each with different laws. To express this idea in figures the theory permit 10^{500}, and only one match up with the kind of universe we know. Then again, if no single theory, no matter how complex it might be, is able to describe every aspect of the universe, then, the universe is of a tremendous complexity, far beyond the possibility for the human brain to create an integral and unique theory to explain it.

Nature things are likely to be simple when appropriately approach, otherwise, we are in front of an unexplainable event of simple nature. But the issue gets worse when physicists realized that the origin of the universe cannot be explained by Einstein's theory of general relativity. Another theory was mandatory, since general relativity does not take into account the particle matter, which is of entire domain of quantum theory. Then, the combination of quantum and relativity theories produced another approach to explain the origin of the universe, another theory that made time-space dimension a little bit more complicated.

Ultimately, we are the outcome of quantum fluctuations at the beginning of the universe. In science, for a theory to be a good theory, it has to be testable. If quantum evaluations on the present universe point to the acceptance of the Big Bang theory, which states that the beginning of the universe started with an enormous concentration of energy into an infinitesimal space. Then, that energy exploded and expanded at a great speed, probably surpassing the speed of light. From all these a question arises: how that enormous amount of energy could be compressed in such a way and in such infinitesimal space to produce the Big Bang? How was this phenomenon performed? Was it out of the blues? What laws or law, if any, can explain such a concentration of incommensurable energy? No theory, holistically, reach a convincing explanation of that immense concentration of energy in such an infinitesimal space. The Big Bang theory is not possible to be testable on a true certainty context, it is rather tested indirectly. It is not possible to re-create such a dimensional phenomenon under lab conditions, not even in a very small-scale setup.

Nowadays, strong evidence is at hand that point to the fact that matter and energy, going through a natural process of increasing complexity, is impulsively switched into new higher organizational condition. It is not coming from pre-existing lower points of energy and matter; both are, in fact, two aspects of the same phenomenon. This is absolutely true when open systems are **thrust away from their common state of equilibrium**. Furthermore, the evidence of such transition to a higher level of organization has been observed in biological systems. This could be a possible reason for the emergence of life itself, and to any other quality of existence, since the law of transition is not limited to matter *per se*. Human beings are open systems. You may apply this law to our own existence, and make your own reasoning, and your very own conclusions; material life, while it endures, is a common state of equilibrium. Furthermore, our perception of the world around is incomplete and not direct since it depends on a sort of lens, which is nothing but the interpretation of our limited human brains. Remember that our brain creates or builds a mental picture of what we see, through a complex process to create a "model" or dummy of all the things we look at. So our brain creates our reality, with innumerable shortcomings, then the reality based on our brain is limited in essence.

The classic Newton's laws of physics certainly describe the phenomena we see around us in our daily life reality. But it is not the only reality around us and in the universe. Then, quantum physics, during the first forty years of the last century, gets on the stage of science as a new model of reality to provide us with another representation of the universe. Again, fundamental forces as electromagnetism and gravity, mentioned before, are included in a new conception scheme. Then, a logical question arises; does this new approach of quantum physics provide plausible explanations in accordance to everyday experience as classic physics theory did? It does, since every single phenomenon and entity in the universe is composed of the same structural unit, atoms; they comply with quantum physics, hence, the macro-world complies as well. According to scholars, quantum physics perfectly matches with scientific observation.

Different theories have been proposed in an attempt to approach "a theory of everything" to unify the four classes of forces in the universe (electromagnetism, gravity, nuclear weak force, and nuclear strong force), then compatible with quantum theory. In one way or the other, those single theories have proved insufficient. It is out of the scope of this book to discuss each of these *in extenso*. Nevertheless, I feel necessary to mention them to provide the reader with the option to look for them, not only in the Recommended Bibliography at the end of this chapter, but anywhere else. The best-known theories are the *supersymmetry*, the *supergravity*, and the *string theory*. The M-theory has been explained already in previous paragraphs as the most plausible approach to understand the beginning and evolution of the universe. But then again, that network of laws is provisional, and future research may impose variations not well foreseen at present. Theories, in the present context, genuinely are trying to explain what is almost unexplainable.

From the simplest to the most complex life forms, the pattern is the same: to be born, grow, reproduce, go through aging, and finally die. Humans are not different, albeit we seem to be the most advanced product of evolution; nothing is different for humans. In fact, humans suffer so much more than any other species during their lifetime on Earth, just because we are rational beings. We understand the processes of illness and aging, the implications and consequences of a hopeless medical diagnosis, and finally, we realize that our final material phase is death. No other animal on Earth is able to understand such a process. Above all, we realize all the consequences and implications that run along the way. Probably, this is the most expensive price we have to afford for being rational beings. Nothing is for free, not even on the miraculous process of life on Earth.

Crocodiles and sharks have been on Earth for millions of years with almost no change; humans just a mere fraction of that time. These species populations have been keeping a perfect ecological balance in their ecosystems albeit both are, in fact, powerful predators, well adapted and extraordinarily skillful. No ecological harm has ever been generated due to the activity of these two species, among many others, dinosaurs included. On the other hand, humans have become a real menace for many ecosystems in our world, up to the point that our planet is in a real danger in many ways, just because humans' activities.

Along human history a fundamental question has arisen. Since the connotation of the answer would change all the dogmas science has sustained as the flag of progress for human knowledge, few, among scientists of all times, dare to approach that simple question: **Are we just biological beings as a direct product of natural selection and evolution and nothing else?** This question has been in the air throughout the book. In this particular chapter evidences, of the available kind, will be discussed and exposed. The reader may apply his/her own understanding and critical reasoning to arrive to certain conclusions. In fact, humankind has been discovering and applying principles, rules, and knowledge according to the progress of human's intellect. The progress and knowledge have been constrained to the evolution of ideas prevailing in time and space, consistently, and in any epoch.

The so-called Great Leap Forward that situated humans as a distinct species is a puzzle in paleoanthropology premises. Most specialists, mainly geneticists, do agree that just a very small portion of our DNA, probably from 0.1 to 0.3 percent of it, made the trick,. The consequences of that minuscule change in DNA sequence are not congruent with the enormous change in itself, together with all the implications put forward in human development. The spoken language acquired by humans, based on proper anatomical changes, has to be the answer. Nevertheless, how could natural selection lead the tiny appropriate genetic changes to produce spoken language? What were the environmental pressures that compelled the genome of ancient humans to develop such anatomical changes strictly to master language? I say environmental pressures just because natural selection acts only and exclusively under the influence of environment, selecting the fittest trait to comply with those environmental requirements. Then again, the anatomical changes necessary to produce spoken language involve the flexibility and proper movement of the tongue, changes in the structure of the larynx, nasal passage, as well as neurological connections to command the complex outcome, of finely tuned vocal sounds, that comprises spoken language.

All these major changes are very difficult, if not impossible, to explain exclusively on a very small genetic change, as discussed on the figures in previous lines. Furthermore, assuming that natural selection did a great job selecting the suitable genes, or complex of genes to produce speech, it is straightforward that those genes already were present in the *Homo* genome; natural selection does not "create" genes, just favors and selects the existing ones. How those magic genes came to be? When and how were they placed in our DNA waiting for the action of natural selection to put them to work? How many mutations took place in the **right sequence** to endow us with such a marvelous gift as language is? We are the only extant species on planet Earth to boast spoken language. Of course, I am not saying that *Homo erectus, Homo heidelbergensis* and Neanderthals were completely deprived of language; it was probably present, though in a more rudimentary form, long before Cro-Magnons appeared on the stage. Those *Homo* species disappeared, and we are here, with full language skills, thousands of languages all over the world, some of which have disappeared from human culture since long ago. Languages also evolve, and a root language may derive to, or give birth to others modified, simplified or even better.

There is another important aspect about language. Primitive extant people do not speak primitive languages; hence, the study of language evolution is quite difficult, since there is not a direct relation between these two occurrences. Besides, linguistic studies have not been accomplished in chimps and bonobos, our closest relatives, at least, not consistently, mainly due to technological and logistic problems since both species move around in an area of several miles. This makes very difficult for researchers to approach and follow them with the necessary equipment to record and study all the vocalizations these apes construct. Yet, the clue how we acquired and developed spoken language remains unknown. Only guessing is possible, but the real fact how complex or primitive language was present in Cro-Magnons, is unclear.

There is a critical point that science, I mean orthodox modern science, try to avoid *a ultranza* for its approach is as difficult as unreachable due to the lack of both wisdom and perception, which were held by ancient civilizations already gone with the rains of time. They were precluded to survive up to modern times due to wars, persecutions, sudden environmental changes, and natural catastrophes. Nowadays, several historians do realize that during a certain period, in the evolution of the ideas, a high quality of wisdom was achieved, and then, lost forever. Furthermore, certain manuscripts have been secluded for dogmatic reasons, since those ancient writings may produce an unleveling effect in belief systems. Neither publications have been issued based on those documents nor the content of such documents mentioned, only in very general terms, if any. On the other hand, science tries to be a special separated part of humankind knowledge, with obstinate stubborn boundaries that recall a popular poster: "Forbidden tresspassing."

But there is a striking contradiction that surpasses any logical approach. Scientists are completely unwilling to move out of the natural selection-evolution exclusive context to explain the presence of humans. Do all scientists hold the philosophical beliefs of materialism? Only if they really do, it is comprehensible that they avoid accepting not even the least contact with other possible explanations, apart from the orthodox science. If they embrace any other belief system different from materialism, how can they outcast such a conviction when dealing with science? Then again, they are converted materialists or they are caught between two worlds. Both are a conceptual and an ideological trap, despite the good judgment of their science. In the family context they praise a divine being, whichever their belief system might be, and in the lab they denied such a trust. If so, this is neither the behavior of a faithful person nor the attitude of a genuine scientist. Probably, what lies behind, is the stereotyped way of thinking that the emerging embryo of modern science, during its appearance along the 16th and 17th centuries, stamped in the human mind as an incontrovertible method to approach and prove the truth.

As any other strict and pragmatic system, science began to lose adequacy to face uncommon events from different approaches, out of the schematic emblem of orthodox science. Science and religion are tangential in this point; both are dogmatic to a great extent. Lessons have been learned that demonstrate that the contribution of different disciplines to approach the most truthful result, is the adequate way. Nevertheless, not all the disciplines are allowed to be together in approaching certain concerns. A very recent and clear example in understanding human evolution and ancestral inheritance, was when other disciplines, as molecular genetics, paleoanthropology, and of course, the pin hold of genome studies teamed together with Anthropology. This latter, alone, could not get further, since the complexity of the events required much more than fossils.

A holistic approach is mandatory in every single subject matter of human concern. Non-discriminatory policy, regarding different disciplines, is the reasonable trend to fulfill the goals in modern research, even in the case when certain approaches are not yet credited as "science" *per se*. Genetics was not a science at the turn of the last century, and now is a leading one.

In the universe everything is connected, all the universe is constituted by the same atoms and molecules, we are part of the universe and science cannot forget that principle. The interconnections are transcendent issues in understanding the universe. The discipline ruled by the strict method of hypotheses testing and reproducibility of phenomena, *sine qua non* any assertion is rejected, is just a method but is not the truth itself. The Big-Bang expansion of the Universe is just a theory, very difficult to prove, and is accepted by many scholars, though. Nowadays, many doubts exist against the validity of this theory. It is based on a rather great amount of hypothetical units never observed before by aligned observations of astronomers. One of its fundamental flaws is that it has been unable to make any quantitative forecast, which has been authenticated by observation. Latest facts indicate that the universe is not the same in all directions and places; the opposite has been a fundamental assumption of the Big Bang theory.

About two centuries ago the concept of *phlogiston* was utterly accepted as a substance included in all matter that burns. This hypothetical substance escaped during combustion leaving but ashes. Today we may deride at such assumption about the condition of burning matter. It sounds like a childish proposition. Were those scientists stupid? No, they were not; it was just the point at which scientific thought was standing at the time. Humankind was just struggling to understand what the world around them was. In the same way, perhaps, scientists in future times may laugh a little at what today scholars do not dare to face. Are nowadays scholars stupid? No, they are not. They are just the product of their time. They cannot be back; they cannot be ahead of their own time. Just few, will dare to leap into extremely new conceptions that will trespass their own time, and into the future, perhaps with one hundred years projection.

Then again, we are the rationals, the up-right walkers, the ultra-socials, the intelligents, the prouds of being, the invaders, the conquerors, the superb family-bonding species, the melody and poem makers, the technology producers, the kings of the air. We are the most dangerous species on Earth. We have to face the truth in its barest implications. Our intelligence and its products, such as technology in all its manifestations, turn out to be the most perilous expansion all over the planet ever seen before. We do not foresee,using our rational gift, what the consequences might be according to our decisions and actions, and if we do, sometimes we do not really care. We may spoil mountains, rivers, forests, gulfs, oceans, archipelagos, islands, even continents, on behalf of social and technological development. We may set aside other human populations, or worst, our own population, just because many humans believe they own the planet.

It is not the industrialized world, which is responsible for catastrophic effects on ecosystems, including the extinction of thousands of animal and plant species. Paleontologists and archeologists have strongly documented important discoveries that shed light on events ocurred thousands of years before the industrialization era appear as a revolutionary social and economic phenomenon. Giant birds, the most famous is the moa, disappeared from New Zealand long before Europeans arrived. Archeologists have found thousands of Moas' bones from roasting Maoris' sites. More than twenty-five bird species have been described based on the unearthed bones. According to radiocarbon dating procedures, all extinct species were there before the ancestors of the Maoris arrived around 1,000 years to present. The aftermath is that within few centuries all those species disappeared. It is not a mere coincidence that after living on the island for millions of years, their extinction happened to occur coincidently after the arrival of Maoris. It is, when the Maoris arrived they found an integral faunistic panorama, settled and thriving, well adapted to the ecosystems of New Zealand. From there on, the process of extinction, in a massive manner, began.

New Zealand was not an atypical phenomenon regarding extinctions carried out by stepping ashore *Homo sapiens*, other Pacific islands, important from a botanical and zoological point of view, suffered similar misfortunes. The Polynesian settlement, all over the Hawaiian archipelago, had the same disastrous effect on many bird species, which have been identified by the unearthed bones, dating some 1,500 years ago. Yet again, the settlement of humans on different places, and mainly in those patches of land as islands are, show a relation with the extinction of many species, whose remains testify their existence in those remote places. Another example is the island of Madagascar, situated some 200 miles to the east of Africa. This island, which has been carefully studied by zoologists, archeologists, and paleoanthropologists was the scenery and grounds of giant birds, huge mammals, and reptiles. Remains of flightless birds, up to ten feet tall and exhibiting a massive build-up of about 1,000 pounds of weigh, were found.

From twelve to thirteen thousand years ago, a non-iced feasible passageway, western of the Rocky Mountains was opened, offering an ideal opportunity to the ancestors of North and South American Indians to go through. These ancient pioneers are known as "the Clovis people", given that the first find of their stone tools was unearthed at the small town of Clovis, New Mexico. Further excavations at Clovis sites demonstrated that many species of big animals were the hunting target of the Clovis, including the enormous mammoth. As in many other places such animal species were not afraid of the human presence, they did not evolved in places where humans roamed. Those ancient Indians found very naïve big mammals, including the mammoth, of course. Such situation made them an easy prey for those early hunter-gatherers. This scenario made possible for Clovis to spread all over the New World in about a thousand years, exterminating many species as they progress in their unstoppable journey to the South. All of the above is well documented based on multi-disciplinary studies and research. Surprisingly, they abruptly disappeared. Some experts suspect that besides other possible explanations, a sustained over-hunting on many important animal species declined the feeding possibilities of these natives, whose population increased rapidly in North and South America as well. The scarcity of game could be the trigger to sweep them off.

All these facts prove that humans did not feel, and do not feel in many cases, incorporated themselves to Nature as an integral element of it. On the contrary, since remote times, as long as specialists can show proofs, humans exert a ruthless use of natural resources, depending on the time and technology at hand. Humans, due to an unexplainable condition, felt and feel themselves as an alien element regarding Nature. They did not, and many of us do not, understand that we are here because we are part of Nature. In any undisturbed ecosystem, there is no predator that may drive its preys to extinction. It is, and has been, a perfect equilibrium in any ecosystem, until the humans entered the stage. The real concern is that extinctions have not been stopped or decrease in numbers ever since. Powerful means of hunting ever increasing put a dead mark on many species as humans, probably since 50,000 years to present, arrived to their premises.

It is not my intention to offer a detail list of extinction events, connected to the arrival of *Homo sapiens* at those unfortunate places, the recommended literature at the end of this chapter is mandatory for those readers really interested to go deeply in the topic, particularly chapters 17 and 18 in the outstanding book written by Jared Diamond. A full and documented description of human presence related to zoological and botanical extinctions, is utterly presented in his work.

We have created what no other species are, or will be, able to do: trading, commercial, and financial activities. Hand by hand with these activities, political and economical interests have been developed, to such a dimension that our quality of life is ever growing farther from our own nature as biological beings. Wars are one of the direct products of political and economical interests, including religious significance as well, ever present since recorded history; the Crusades are a well recorded historical example of this.

Our genetic design does not match with most of the foods and liquids we consume. We do not socialize with others, not enough as it is supposed to be the rule for social beings as we are. The human's solitude is evidently increasing with technology, called the modern life-style. We became different from the apes, to some extent, for our long legs are designed for long upright walking. Cars are so comfortable that more than sixty percent of our walking necessary activity, has been cut-off. We are a product of many factors in modern times, but unfortunately, not all those factors are healthy for us. When I say modern times, I mean the lapse since the end of medieval times on; that is when mankind started a rapid increasing progress regarding inventions and technology, sadly including war technology.

We have acquired deleterious habits as alcohol, smoking, and drugs, all menacing our organisms and families as well, including our healthy future in terms of humankind. Just compare the addiction possibilities for apes and monkeys in the middle of the jungle where an uncountable number of plant species are at hand that may cause drug-effect outcomes. Up to present, there is not even the slightest evidence that apes or monkeys ever used such plants to enter an altered-mood condition, albeit they are not rational beings, unable to decide if it is good or not for their health. I speculate that some kind of evolutionary "preventive-alarm" keep them all away from such botanical substances. Perhaps we have lost such an "evolutionary alarm" on behalf of our rational thought; it sounds so paradoxical. Then, we are rational just to some extent. We are unable to be fully rational if we attempt against our own beings, using and spreading dangerous substances that annihilate the life of many human beings, in a short or in a long term basis. As long as I know, no animal species ever intent any action against their survival. On the contrary, animals show a rapid response to preserve life in front of any odd smell or taste.

It looks like the pace of evolution has not been swift enough to time with the advancements of technology, which demand certain biological fitness, not present in modern humans. An unquestionable evidence is the spread of anxiety disorders, depression, and panic attacks, utterly common in modern times. Our psychological setup is not fitted or design for such demanding situations humans are exposed to, due to our "modern lifestyle."

The transit from Cro-magnon times to present has been too speedy for an appropriate evolutionary process to take place. Our biological performace is not in accordance with the rapid technology take-over on our daily lives. Inventions go at the forefront compare to our biological possibilities in coping with such tremendous progress. Our own intellectual development has been in conflict with our psycho-physiological possibilities.

Stress is a modern word that try to name a new form of incapacity to cope with the fleetness of a daily living style. Our minds are not at the proper level required by the imminent future world, in which a more hurried living mode will settle down. The question that may arise is: up to what extreme pace future inventions and technology would be harmful for the model of humans we are? Then again, there is an incompatible situation between human evolution and the unstoppable progress of top of the line technology.

It is not very difficult to infer that addictions are a by-product of rationality. Unfortunately, to be rational is a double-edge sword, you may be seriously hurt by your own decisions, even worst, you may hurt others that are not responsible for your wrong actions. We are free to decide what way to take in the middle of the crossroad. Fortunately, or unfortunately, every single human being is unique, though many rational decisions coincide on a collective basis as has been demonstrated in the case of wrong rulers along the history of political and social problems; nazism, fascism and communism are modern examples of such deviations from the well being of people. A group of ill-thinking adepts coincided in time and space to impose their thoughts and doings to a whole country and beyond. Such happenings always occurred under very particular economical, social, and political conditions.

At the end, those deviations stumbled down and disappeared from the human society context, as medieval times and slavery did. Aberrations never endure, their counter-part is the whole planet population, that, in one way or another reach a social and economical balance to go ahead and thrive. Farmers, nowadays, are living in excellent conditions and commodities compare to XVI century ones. Even more, at present many farmers around the world are wealthy. That was completely impossible 300 years ago. Of course, farmers wealthiness not only depend on their own struggling effort, it depends on the fairness and good judgment of honest and capable governments as well. Unfortunately, not all countries around the world can count on this.

One aspect is worth to mention, lifespan in humans is not exactly in right accordance with the high developed organisms that we represent. Some species like the Galapagos tortoise may live up to more than 140 years, the Koi fish so popular in private ponds and aquariums have been recorded very close to 200 years. On the other hand, the top lifespan in humans never exceeded 122 years. Are these fish and tortoise DNA more perfect than humans'? Beyond that, the jellyfish *Turritopsis nutricula*, a very low organism in the animal kingdom is practically immortal, they are capable of processing regeneration on all its organs and tissues, without ending. Then again, why a tortoise, a fresh water fish, and a jellyfish are top of the line in life span and immortality? In multicellular organisms, as we are, cells are frequently damaged and repairing processes have to be prompted to avoid serious biological problems. Aging is a process where the repairing mechanisms are not so efficient, futhermore, many times they failed to fulfill its commitments.This may lead to an organic disease, or worst, to an increase in damaged cells un-controlled division and multiplication, turning into a neoplastic event, tumors, benign or malignant. Other animals, mentioned before, are capable of maintaining repairing processes for a long time, hence aging progression is significantly delayed and lifespan considerably extended. Their DNA are endowed with enough genetic information to secure the cell repairing process much better than ours.

In those organisms two possible processes, with high efficiency, may occur to avoid serious damage to the individual: **apoptosis**, in which the damaged cell kills itself to prevent abnormal reproduction that may lead to anomalous tissues, neoplastic process. The other preventive course is known as **replicative senescence**, in which the cell simply stop division, hence proliferation of abnormal tissues as well. It is not very hard to infer that humans, as a majority, do not posses such efficient processes since many diseases emerge in young people and children, too. Besides, genetic-based diseases are listed in huge numbers, albeit some of them are also connected to environmental problems very difficult to detect, not only their kind but also the level of impact in humans' biology weakness. At present, about 800 genetic-based diseases are recognized. Expectations point to the possibility more shall be enlisted in years to come.

It is worth to mention that some scholars discuss on the reason why we are not immortals, or why aging is inevitable. That is not exactly the point of interest, one thing is the unavoidable aging process and another very different is the proper defense mechanisms against so many diseases that we suffer during our lifetime. It is very simple, our apoptosis and replicative senescence mechanisms are entirely imperfect or not finished at all, since natural selection stopped to act upon, or is still acting but at a very slow pace. Though statistics are not available to make comparisons with human populations 10,000 years ago, different kinds of diseases are appearing *de novo*. I personally know the case of a close friend of mine that a medical team working in a modern hospital in the United States, declared they were incompetent to diagnose the medical condition of my friend; it was something new, not described in medicine books, or treated before in any medical facility.

As I mentioned before in a previous chapter, a great amount of energy is required to maintain our brains in top conditions, we have to pay a high biological fee for our unique intellect. Perhaps, that high requirement of energy our brain demands, prevent the rest of the cells in our body to use enough energy for apoptosis and replicative senescence, or we definitively do not posses such mechanisms in an efficient manner as other animals do. Then, we have to face the incontrovertible truth that natural selection did not complete its job on us, or we have not reached the utmost condition of "perfection" as the most adapted *Homo* species on Earth, though our presence removed Neanderthals from Earth, or so believe.

It has been stated by some specialists that natural selection, and then evolution, have disposed biological processes in such a way that all our systems decline under certain arrangement, comprising physiological and organic complications ending up with a whole deterioration of our body. Some examples include cardiac problems, fragility of bones, circulatory disturbances, inefficient kidney function, just to mention a few. According to these specialists, natural selection deters the energy investment of repairing organs and systems when the reproductive cycle, which guarantees the perpetuation of the species, is declining. There is no purpose to such energy investment if there is no offspring to add to the next generation. It is, natural selection must match deterioration in all physiological systems as aging progresses. It does not sound very clear that natural selection acts according to the aforementioned actions and is unable to improve our DNA, which lack, as already insisted somewhere before, the biochemical elements to go on with proper repairs to improve our life span and health condition. To destroy such theory of energy investment based on fertility and offspring production, we have the examples of young people, including children, suffering from diseases such as nephritis, cancer, cardiac conditions, liver malfunctions and many others. These diseases, in young human beings, are precluding the offspring for the next generation to be, then deterioration of biological systems not only and not exclusively occur in unable aged reproductive individuals. That is not the point; the culprit is the extreme imperfection of our DNA that indiscriminately deteriorates our biological systems to death, no matter how old we are. It is a human condition, with an unknown answer to convince the more skeptical researchers. If we take a closer look over the huge biodiversity on our planet, we will find that life span is typical for every single species, on the average; *Homo sapiens* is rather an exception. Only those species that have been genetically modified by the intervention of humans are subject to abrupt changes in their life span mean value.

The fact that we enjoy many technological advances does not mean that we are a fully developed species, perhaps a long evolutionary journey still lies ahead. We are not the culmination or the finished product of evolutionary changes to be completely humans in the most entire meaning of the word. Profound discrepancies and dangerous issues still exist among different countries, we do not treat ourselves as one and unique species, we see other humans as different.

Unfortunately, there are rulers that seize the right to mislead their countries in a very unsafe confrontational scenario. We are all alike; political, religious, economical or any other sort of circumstances must not put countries apart as enemies, which usually lead to warfare. We are a problematic species, in such a degree that we attack to death other members of our own species. This is not the rule in the animal kingdom, only chimpanzees show a similar, but less aggressive, behavior against other groups of the same species. We have to accept that our aggressive actions have been all along the way in our evolutionary process; we are nothing but evolved apes, though it is hard for many people to accept it. This is our true genetic background and our ancestral drawback.

Not even the most powerful predators ever use the same prey-killing techniques when facing a dispute with members of their kind. The repertoires for menacing opponents of the same species never include the deadly weapons they use in hunting the prey. If such misconduct is carried out, the species itself may be in danger. At the end, it tends to disappear. Natural selection has provided those powerful predators with inhibitions as meaningful mechanisms to avoid such destructive actions.

Extinction is the price for an aberrant conduct used against members of their kind. *Homo sapiens* is the only extant species on the planet that dare, with no restrictions, to launch deadly attacks against members of its own species; not only on an individual basis, but against certain groups and populations as well. Sadly, in fact, there are not so uncommon cases where a member of a family seriously hurt or kill another member.

Genocide is another drawback tightly found in our behavior. Since written history, these happenings have appeared in all epochs. We have many examples pictured by history: the Roman War against Carthage, the epics of Greeks and Trojans, European conquerors devastating the New World Indians and Australian Aborigines. All these happened with a common ending: the slaughter of people irrespective of age or sex, involved or not in the defense of their legitimate territory.

Last century, the Argentine military government carried out bloodiest genocides, shaped with an unprecedented cruelty. In less than a decade more than 10,000 people were counted as *desaparecidos*, comprising opponents and their families, including women and children. Of course, we have to add that Hitler commanded massacres on Jews in the concentration camps. Those committed by the Union of Soviet Socialist Republics government from 1917 to 1959 adding up to more than sixty-six million people, just because they were opponents. Humans killing humans, in such an amount that no beast in the jungle ever approaches such a killing record, not even on their preys.

Despite such evidences there are specialists that still seem to be reluctant to accept that we are an extremely aggressive species, and try to soften the reality invoking social and cultural attributes as the real culprits. This pretense is nothing but a gross euphemism.

We descend from beasts, though natural selection has favored altruism, benevolence, cooperation, understanding, charity, compassion, love and true friendship as the most advance human virtues. Unfortunately, some remnants of our bestial common ancestor still have a secluded place in some members of our species. I do not deny the well-known and universal formula that **phenotype = genotype + environment.** Precisely, certain unknown genes might be responsible for such aberrant behaviors. These are unknown at present, though the human genome is a formidable tool in the hands of geneticists. Perhaps in the future, these and other unpleasant traits could be detected for the sake of future generations. I speculate that the mere appearance of *Homo* on earth carried the imprinting of such aggressive exclusive behavior. Nowadays, many are the cases, well documented, that individuals raised in a family where both violence and aggressiveness were utterly common, run off their families. This demonstrates that the environment cannot be invoked as the only responsible factor for misbehavior. Those individuals did not show up those patterns, they fled instead.

Human behavior is conceived according to natural laws, but at the same time, it is so a complex phenomenon. To really determine its complete outcome science has to take into account innumerable variables, hence, predictions turn into impossible. According to a holistic approach, one has to consider a dynamic complex of trillions of cells in the human body, which is impossible for present science knowledge, which in fact would take an unpredictable length of time, probably beyond our own existence as species. Models have been created to figure out the intrinsic mechanism of humans' behavior. No one has been accurate enough to explain all possible variations, just general concepts and approaches have been accomplished.

We lack a great deal of excellence to really consider ourselves a natural selection gifted- species. Nonetheless, we cannot overlook other possible answers as explanations for such a contrasting and apparent imperfection compare to lower animals in the zoological scale. To approach such a solution, or at least to state some plausible hypotheses, scholars ought to open their minds to any possible research though it may seem intricate and out of context. No effort must be marginated. Open minded scientists will approach in the future what today is an avoided path that seems so distant from science.

In fact, the limits of science are nothing but the limits and perceptions of humans who are the architects of the theoretical and practical discipline we call science. Just take a look at the concepts of any branch of the scientific thought 100 years ago, and you will find how deeply they differ from contemporary approaches and reasoning. We cannot blame those scientists, at the beginning of the last century, for they short seeing and flaws. To become aware and convinced of the right approach is just a matter of time, as well as natural selection and evolutions are compelled to time. What may seem untouchable, and even more, spurious for the orthodox scientific thoughts in our time, could be a new and undeniable approach in years to come. A hard and difficult road may stretch out ahead but at the end of it, wisdom is the close encounter scientists are looking for.

The Essence

An undoubted event is the inclusion of meat in *Homo* diet. This went to great length in the posterior evolution of humans, in the widest meaning of this occurrence, since the ancestral forms of apes were fruit and vegetables eaters with just a low proportion of small prey included in their diet. Along thousands of years, the time can hardly be determined, of consistent evolution and adaptation to new environmental conditions, our ancestral relatives became carnivores. This occurred in a very special and unique way to be so, since genuine carnivores have powerful teeth and claws, which they use as super weapons. These structures are of such tremendous danger in any confrontational event and social antagonist encounters that the creatures involved in, could be killed or seriously hurt. In some way, natural selection has favored powerful inhibition mechanisms to prevent such encounters to take place; the survival of the species involved could be seriously endangered otherwise.

On the other hand, killing is not an important task for primates (remember we are primates) since they are a fruit-picking species though they eat some insects and chicks from any nest not well guarded. Small mammals included if they have the chance to get a hold on them. We are nothing but a very specialized and evolved kind of ape. We directly descend from fruit and vegetable-eating apes, our teeth are a strong proof of this fact. On the other hand, we became hunters, the first hunting-ape, albeit still gathering fruits from the trees as part of our natural diet.

To hunt means to be a carnivore, hunting has no meaning otherwise. This combination is what made us hunter-gatherers. This unusual combination involves something difficult to explain since the anatomy, in all its features, of hunters is very different from the anatomy of gatherers as monkeys and apes are. Carnivores as tigers, leopards, lions, wolves, foxes, dogs, and hyenas are a clear example of what a genuine hunter (predator) is expected to be.

There is not a simple explanation to approach the fact that a primate, as we are, millions of years ago became a hunter without any of the anatomical and physiological features inherent to carnivores, predators understood. We are the ultimate exception of a hunter-primate, no other fossil primates attest for such possibility. Other hominid forms, as *Australopithecus* were too small and lightweight to attempt hunting as a means of subsistence. *Homo* is the exceptional being, weaker than lions and hyenas, so much less speedy and slower in the sprint than those well-bestowed predators, not even a four-leg runner.

How could this odd model of carnivore succeed in the long evolutionary pathway without any of the top requirements to be a genuine hunter? The answer is "brain," which was different from, and much more developed, than any other mammal's. Lions are powerful predators, armed with deadly teeth and paws, which use to rip apart and smash down their prey; *Homo* could kill from a distance using hand-made weapons, spears and arrows. A high-developed brain substituted the potent carnivore's weapons.

The answer is just "brain," but the real question is how this came to be such a developed organ, while in other mammals changes are not substantial, not even in apes. Hyenas have been performing the same behavioral pattern for hundreds of thousand years with no apparent modifications. The study of human genome indicates that there are not examples of fast natural selection action. On the contrary, most of the activity of natural selection, noticeable in the genome, seems to have occurred during tens of thousand years, perhaps more. How changes in the human's brain took place up to the point where we are today? Probably, that point was not very different from 30,000 years ago Cro-Magnons. It is the core of the big incognita about human race.

Furthermore, today we know that only 2% of the human genome synthesize proteins, and roughly another 2% is engaged in the regulation of genes. Again, the action of regulatory genes is to determine in what quantity and how well those proteins will be produced. When we say proteins we are talking about the synthesis of tissues and organs, since proteins are the raw material of cells. The rest of the genome remains unknown, no evidences exist what its role is.

What are they there for? Are just an immense amount of genes with nothing to do from a biological point of view? Is it possible that natural selection overlooked such a genetic potential through millions of years since the first hominids came to be? It sounds absurd from the very moment that we conferred a tremendous selective power to natural selection. Or, is it that natural selection has nothing to do with this enormous genetic potential of non-working genes? There is no answer for the moment time. Science is in complete ignorance about that substantial portion of the human genome that do nothing; millions of genes with no apparent purpose.

Little do we know. But the worst, at this moment, is that we do not have any technological means to go further in the investigation of such portion of our genome to inquire about its real significance. It is ridiculous to think, or accept, that such a huge proportion of the human genome is not working just for the fun of it. There is something that today's science is unable to get a hold on. But most scientists simply declare, as a fact, that such colossal part of the genome has no identified role. Are those millions nucleotides (see chapter 1 as a reference) a genetic reserve that will start-up a specific synthetic or encoding task according to environmental pressures in the unpredictable future to come? Are they there just waiting for the exact moment to become active and launch an evolutionary progress, fraction by fraction, according to specific circumstances in future environments? If it is the case, then, we are much more than just the product of natural selection and evolution, since we have a genetic reserve not created by natural selection and not connected to evolutionary processes. Given that, to be selected, any trait has to be favored based on its adaptive functionality. Those millions of genes on that portion of the genome, do nothing, hence they have not been exposed to the action of natural selection. They exist, they are there independently of natural selection. The existence of this enormous amount of silent genes demonstrates that the onset and long lasting existence of genes, not always is connected to the action of natural selection. How a complex organism as ours is could fully function with only roughly 4% of its genetic potential? It sounds as a waste of biochemical energy to sustain that huge amount of genes with nothing to do. It is really sad that science be unable to approach an explanation and then a proof to actually understand the meaning of such a genetic phenomenon in our own species. How could be possible that such amount of genetic codification has no a definite purpose? Are they dormant genes to be activated upon upcoming changes in diverse human population conditions? I would like to expose some figures resulting from previous studies accomplished several years ago.

First, around 1690 the world human population was about 495 to 510 million people. At present, the number is not less than 7 billion, and every single day it increases from 145,000 to 155,000 new people all around the world. This will lead us to an overcrowded planet in just about 250 years from now. The resulting population density would be more than 10,000 people per square mile. Other calculations launched the prediction that human population is doubling-up every forty years or so. Recent calculations by world institutions predict around 10 billion people living on Earth by the year 2050. All these are rough calculations based on statistics accumulated during many previous years, but the real outcome could be much larger. This prognosis does not come from any particular calendar or esoteric predictions. These are simple scientific based statistical figures which allow the specialists to make projections for the future to come. The explosion of population growth in China and India is alarming. Besides, the daily living conditions for the people in those countries and other developing countries are not the best. The planet will be hopelessly crowded in the next two centuries or so.

The worst, some calculations foresee that according to present oil increasing consumption rates, this resource will be exhausted in the next 40 years; a true catastrophic event with unpredictable consequences for humankind. Developing technology in ground transportation, air transportation, electric power, and industrial use, are fundamentally based on oil as a resource.

It is difficult to stop the pace of an increasing dynamic population, and an overcrowded planet is a real menace for our own species, and for many others. This population phenomenon delineates the "beginning of the end" for *Homo sapiens,* as it could be for any other species under the same circumstances. As mentioned before, no species is forever, ours is not an exception. Devastation of our planet has started since long ago. Many natural areas have been destroyed due to industrial and economic purposes, as a consequence, many animal species have been sent to extinction or in the brink of extinction in all continents. The global warming is one of the consequences of our destructive activities, and the recent oil spill in the Gulf of Mexico is a palpable proof of our poor control on technology. Our technological inventions represent an icon of progress but at the same time they may become lethal weapons against nature; examples abound. We do not have a true control of our technological endeavor, much of which, at the end, burst out as dangerous activities against our common home, planet Earth.

The food shortage in the world is another fact that menaces the partial extinction of certain populations at present. Not all human beings have the basic access to a fairly balance nutrition. We are many and the planet resources are every year scarcer.

Soon, many of the natural areas will be completely populated by humans, in a despair effort to expand the limits of overcrowded cities, looking for better places to live in. No species in the world can tolerate an overcrowding phenomenon without the perilous consequences of a near extinction shadow. Then again, are we endowed, from a genetic point of view, to face and survive and overcrowded condition? Are those quiescent genes the biological reserve to preserve the species in spite of an overcrowded planet? Will they definitely separate the survivors from the rest based on the crude action of natural selection? There is no answer by the moment time. Nevertheless, we have to analyze the evolutionary process of *Homo sapiens* in particular, as an attempt to understand all our accomplishments and all our failures. We are nothing but the product, as descendants, from an unknown form of ape. Though our evolutionary process has been a long journey, it has not been long enough to delete certain genetic background that may still remain to hamper a congruent and consistent behavior as true human beings, aimed to the safety of the species. This sort of behavior is desirable to avoid erroneous actions; our history has many. They have been the common denominator along the known olden times of humankind.

During our evolutionary pathway, some nonsense contradictions appear as unexplainable facts; few paleoanthropologists attempt to explain them with reasonable eloquence. Specialists do agree that our first explicit evidence of art was around 40,000 years ago detected as cave paintings at Lascaux, France, performed by Cro-Magnons. It has been a controversial issue to consider art as a product of natural selection since it does not confer any advantage as transmitting genes to the next generation. How was art onset on human groups? Art, in fact, is the product of abstraction, and the outcome of it is manifested in paintings, sculptures, music, songs, architecture, etc. Was there, 40,000 years ago, any advantage for art performers to get a mate? We will never know, since DNA and bones cannot tell.

Then again, abstract thinking, as it is required to develop art, is an undeniable proof of a superior intellect. In contrast, agriculture and herding showed up in human groups only 10,000 years ago. All specialists coincide with such dating. This is a huge contradiction. Why people so intellect-developed needed 30,000 years to think of and carry out agriculture and herding, which are simple methods compare to painting and sculpture? This means that they were developing art while being simple hunter-gatherers. It sounds ridiculous or nonsense at all. There is an odd gap between Lascaux paintings and the beginning of agriculture. It is too long a time for humans to realize, with a well developed intellect, the need for agriculture and herding that were at hand, given that human groups already enjoyed a social organization, superior to that of Neanderthals.

The point is that "art" is difficult to explain as a natural selection favored trait, since it does not guarantee the transmission of genes from one generation to the other. We cannot project the concept of "an artist" at those times as a Hollywood star, acclaimed by the fans and able to select among the most beautiful *feminas* in town, get married with a wonderful healthy lady and pass genes to the next generation. I doubt that such criteria could exist among cave people. We are a little bit prone to approach ancient human beings according to nowadays-cultural practices. The advent of art is somehow difficult to explain, and then again, natural selection does not have anything to do with such an abstract concept and inspiration, as art is. Then, how was art onset on human groups without the intervention of natural selection? Try to guess again.

Furthermore, many specialists state the fact that people started independently to domesticate animals and grow plants in different parts of the world. A very simple and mandatory question arises: how could they started "independently" to raise plants and animals as a global and almost simultaneous event if they did not have, at the time, any means of communication to share such advances? Ten thousand years ago the planet was under-populated, and to transmit experiences from place to place was not a simple task. Besides, at the time, different languages already were developed and translation ability was not a part time job. How could all this happen in a historical short round in different parts of the world? As I said before, no convincing explanation is around the corner. The bottom line is, as facts demonstrate, hunter-gatherer-artists 30,000 years ago, then farmer-herder- artists from 10,000 years to present.

According to experts, the spread of agriculture and herding mainly occurred in Europe and Eurasia. From there, animal species and plants were appropriate for such activities. In Australia, North and South America, and Africa there were no *ad hoc* animals or plants to develop these two important human practices. Nevertheless, scholars do agree that the dog was probably the first species to become tamed, some 15,000 years ago, followed by sheep and goats some 11,000 years ago; pigs, cows and cats some 10,000 to 10,500 years ago. The horse, probably 5,500 years to present.

Some few scholars try to explain many other human activities from the point of view that agriculture and herding created the course of many other modern activities. They attempt to explain that, from the westernmost part of Europe to the extreme portion of Eurasia, the movement of crops and herding animals was not a problem, since similar climatic conditions predominantly occurred all over that route. The expansion of agricultural-herding practices was blessed by geographic characteristics. This may explain why these practices flourished there and not in Africa, Australia, North and South America, leaving these portions of the world far behind civilization progress. Experts explain that moving from South to North America imply a drastic change in climatic conditions not found in the aforementioned European-Eurasian route. Those few animals and plants that could be grown under human control did not stand such climatic extremes if moving northward. It is perfectly understood, and I do agree with this hypothesis. As a biologist I acknowledge the consequences of transplanting species from one hemisphere to another, mainly in those ancient times where acclimatization of exotics was not in the hands of humans.

The controversial point is that, as mentioned before, all other human progresses according to some specialists rest on agriculture and herding. These activities were successfully based on geography and cultural characteristics in Europe and Eurasia. Let us begin with housing development. It is not well understood why African populations never developed any house modeling beyond rudimentary huts, though Africa is a huge continent full of timber. They never attempted the use of lumber to improve their housing commodities and privacy, not even the simplest models of log cabins have ever been found in the continent. This is, in many aspects, the same and old story as the Stone Age, which lingered for millions of years with no apparent improvement.

Consequently, if woodcraft did not develop for housing improvement, watercraft either. In my very own opinion, I do not see the relation between the use of wood as a construction material and the advancement of agriculture and herding, as some experts try to make a connection between both. Some specialists argue that due to the hot climate in Africa, simple huts were enough for people to dwell. It is to ignore the heavy continental rains that occur in the tropical jungle, which are devastating for those simple huts.

Probably, to be nomadic hunter-gatherers was the only possible way of living without woodcraft development. It is necessary to mention the huge coastal line of the African continent; besides, big rivers abound all over the continent and important lakes as well. Why simple watercrafts were not built as a means of rapid transportation? Tribes necessarily lived along riversides and lake coasts to have access to a safe water supply. The answers to all these questions still remain unknown, since there are no proofs of differences among people inventiveness, as has been mentioned by different scholars.

Another incognita, not solved in our days, is the flourishing civilization of Mayan Indians around two thousand years ago in Central America. They stay outside the theory of agriculture and herding to explain progress in human populations, since they developed agriculture in spite of non-proper plant species for such practices. The answer is not simple either, since they accumulated a huge amount of knowledge. Their achievements included extraordinary architectural designs, astronomy, an outstanding calendar, and a social and cultural organization far beyond their own time. All this was developed in the heart of tropical rain forest. Archeologists understand such progress up to certain logical limits, not beyond, the rest is smart guessing.

The New Extended Knowledge

The integrative work of Darwin, which comprised not only his keen observations on natural science, but also a tremendous amount of scientific information at his time, including the experimental outcomes of breeding domestic animals, was an essential element to build up the well-known Modern Synthesis during 1930's and 1940's. In fact, Modern Synthesis put together pioneering advances from Mendelian knowledge adjoining the advances of quantitative genetics and population genetics. The latter is considered the math-theoretical vertebral column of all evolutionary biology at present. In this way, an enhanced evolutionary theory was born, where the new science of genetics added a remarkable impact on modern biology. Hence, Modern Synthesis came to be the yardstick to measure all other progresses in evolutionary theory. Nevertheless, nowadays-evolutionary theory embraces a great amount of concepts that were not considered on the prime frame of the Modern Synthesis, some seventy years ago.

In modern biological science, the recognition that Modern Synthesis concentrates all its theoretical frame exclusively on natural selection acting upon organismal structures and characters, brought the narrow concept that any direction of evolutionary changes were exclusively the undisputable result of natural selection. All studies and researches suggest that well-established properties of any organism tend to help certain kinds of changes rather than others; then, natural selection is solely a steady condition operating at the surroundings. It comes that, living things determine, in one way or another, the variations to be selected as well as any novelty. Then again, Darwin in his 1859 Origin of

Species...explained that he used the term "chance variation" just because he faced his own ignorance on the cause of variation. It makes clear that Darwin felt himself unable to explain, not only the origin of variation but also its possible directionality.

All along the anthropology and paleoanthropology course of research, human evolution keeps the features of a major transition. Additionally, many experts consider it was an exceptional event, only once in the realm of primates. This conception of the human evolutionary process as a major event is really new, as new that it gives way to other inferences to be tested. Less than twenty years ago evolutionary anthropology and evolutionary psychology began to gain credit into the contextual frame of human evolutionary process. In addition, proliferation of whole genome sequencing technology propagates at such rate, that at the end of 2007 more than 150 genomes sequences, on different species, were completed or in progress. Genomics, as it is known today, facilitate technologies to go deeply in genome structure and function, including many species. Nowadays, it is well understood that the genome is a kind of evolutionary mosaic, which is the result of multi-segment bricks built-up jointly, from different species and families along the extraordinary wide-ranging evolutionary journey, in the history of life. Those "bricks" are known as **transposable elements,** which are the very basis of the genome dynamism. They are, in fact, mobile elements that excise and then insert themselves in different sites along the genome, DNA chain. In humans, which are our target in this chapter, these transposable elements encompass a vast fraction, up to 44% of the whole genome.

Which is the significance of all these transposable elements from an evolutionary point of view? Very precise genetic studies evidenced that insertion of transposable elements are decidedly deleterious, generating certain disturbances in the function of alleles, but at the same time they can lead to novel genes or new gene functions. On the other hand, the real biological phenomenon is that genes do not evolve in seclusion; they evolve in the whole context of the genome. Many appealing evolutionary events are apparently only possible if considered on a large scale, comprising hundreds or thousands genes. All these events seem to be fine-tuned, in such a way that their sequences as well as their appearance are exclusive of a well-coordinated phenomenon. This occurs without the slightest tendency to produce catastrophic effects in the genome, on the contrary, it points to an original event in the evolutionary process, otherwise, very difficult to accomplish.

More than one specialist has affirmed that reiteration and convergence of characters are clear attempts to complete certain structures, or even behavioral traits. According to those scholars, these phenomena are an undisputable and evident recurrent tendency showing up in response to environmental pressures, which indicate that such biological outcomes are inevitable. They are direct results of natural selection and evolution, pertaining to huge lapses of billions of years.

Nonetheless, other plausible interpretation is that, all those features already existed in the genetic endowment of different life forms, or zoological groups, due to a common ancestor inheritance. It was probably in the form of a primordial gene-pack, or transposable elements. Then, as diversification expanded, each particular group showed what seemed to be "convergence," when in fact, it was nothing but the expected development of an ancestral gene pool, emerging in different taxa, at different times, in the evolutionary history of life forms.

At first glance, it could seem to be as an authentic case of convergence, when indeed, it is not at all. The genuine convergence is the appearance *de novo* of the same trait in different groups, or different taxa. Furthermore, why chimps and bonobos did not converge to a more human-like form if they are so closely related to us? We are very much the same. Is it just because we split off just about 6 to 8 million years ago? Do they need a lapse comprising more millions of years to really approach humanness? We will never know. Human existence will not last for millions of years to see the "convergence," if any. Which theoretical approach is the most reasonable? Is it the urging of genetic traits under environmental pressures to converge into certain characters? Or, is it that if the proper gene-pack for those characters is not present, the so-called convergence shall never be? Perhaps chimps and bonobos do not possess that fundamental gene-pack, or transposable elements, to really accomplish the uppermost approach to humanness.

Beyond Biology

Life is energy, and that kind of energy to sustain life comes, essentially, from our foods.All the intricacies of the metabolic process are capable to extract all fundamental nutrients from the food we ingest; almost everybody in the 21st century is convinced of such a simple explanation. If you do not eat, your biological system will collapse in a matter of days, the immediate result is death. Then, physiological pathways are responsible to provide nutrients to every single cell of our body through a very special carrier, the circulatory system. From those processed nutrients dissolved in the blood, organs and tissues obtain the proper energy to develop their tasks, all integrated, in the magnificent mechanism of our biological organism. In our own body we use different kinds of energy, such as mechanical muscle energy to move our skeletal structure. The heart is a continuous-moving muscular complex that needs a constant supply of energy, not to stop. Then, nervous energy is used in all the hyper-complex neurological system, in fact, to be what we are.

Now, a very simple question arises: do humans know all types of energy in the universe? Certainly not. We may recall that, just one hundred years ago, atomic energy was not in the possibilities of technology to be developed; nowadays, it is a fact. Then again, there are many energy sources. Plants obtain their energy from sunlight, carrying on the photosynthesis process. We cannot do that, for we are another kind of life form. Nevertheless, we need exposure to sunlight to elaborate vitamin D, hence, to facilitate the absorption of calcium that we may consume in our regular diet. Fish can obtain oxygen dissolved in the water for them to breath, we cannot. All of the above is Biology, simple and basic, except atomic energy, which is Physics.

How many other energy sources, or energy systems, do exist that we are not aware of? Furthermore, the universe is energy, an uncountable amount of energy in constant change and expansion, developing new systems and destroying others, but the energy is always there. It is difficult to imagine the hugeness of such processes in the interstellar space. The simplest corollary we can obtain is that energy, in itself, is and endless entity of possibilities, the largest part of it completely unknown to extant human societies, and science. Here is where a different knowledge journey starts by itself. Then again, are we exclusively biological beings that sustain life based on food ingestion to make our biosystem work during a very small lifetime? To approach the answer is difficult in the light of orthodox science, under canonical schemes, and without reverence to scientific establishment. To get such an approach, researchers, and there are several, have to face acerbic critics, perhaps untenable positions as researchers, and of course, a cut-off on grants. Nevertheless, in science, as in any other human effort and kindness, there are people ready to face the truth, at least to move toward the truth, regardless of colleagues' derisive comments.

The evidences of a certain unknown energy, linked to human existence, was known at the beginning of writing, by 1600 BC, but probably such experiences dated long before that time. On a rigorous basis, we have to adjoin to the written evidences since any verbal communications have been lost in time. It was not until 1,100 BC that Phoenicians created the so-called alphabetic script, which became the essentials for the development of all modern European scripts. It is uncertain what the first document was to attest those experiences, not related to common human's perceptions. In any case, religious beliefs started, in someway, based on incomprehensible facts, and then adjudicated to an unknown powerful entity, in absence of a better knowledge. Some authors affirm that primitive belief systems were not related to any particular deity; they were only based on the unity, tolerance, help and understanding of all other human beings. In fact, the interpretation of those ancient manuscripts is not always simple to accomplish, and in many instances the specialists have to decipher the texts with missing pieces, which is detrimental for an ample understanding of the message itself. Our civilization is so new that the ten number arithmetic notation, which we use plenty enough, dates back some 1,300 years to present.

As writing developed in different countries, testimonies were printed to let others know about particular experiences certain individuals went through. Both, tall tales and poor detailed experiences, due to inhibited persons, probably appeared in the literature at the time. Besides, second hand stories were heard as well. It was at the beginning a true confusion, between reality and something that escaped from the logical judgment. Many religions tried to refute such experiences, up to the point that many people were accused of heresy. Belief systems pre-established dogmas and everything or event that could haze, at least, a measly aspect of the religious establishment, was condemned.

This has been so a common occurrence that at the end of the 13th century a Bishop of Paris (Tempier), under the instructions of Pope John XXI, released a list of more than 200 condemned heresies; one of these was the unaccepted idea that nature may follow laws, since its acceptance conflicts with the all powerful God. These people were so fundamentalists that not even foresaw the possibility that "nature laws" could be direct God's decisions or will, or even the laws that God decided to act upon Earth, his own creation. Undoubtedly, medieval times were characterized by the most limited and obscure reasoning, preventing not even the advancement of science but the improvement of the most elementary conditions of human feelings and perceptions.

To get it worse, many interpretations of those personal experiences flourished among people and in different countries. At the time, it was impossible to arrive to a consensus to better understand what the phenomenon was. Spirituality was the term that seemed to identify such happenings, but then again, this simple term took different interpretations according to cultural backgrounds and belief systems. When print paper was finally a possible technology, books dealing with the topic of spirituality were issued and profusely distributed. People could not get rid of the primeval confusion either, since personal interpretations, of the same occurrence, were focused and published with different approaches according to the author. For the worse, science was absolutely not in the capacity to provide an elemental contribution to help through a logical approach, in facing the factual evidences.

The pressures of religious dogmas kept on going against the incontrovertible evidences. It was not possible that people from almost all countries of the world were dishonest, making-up stories as propaganda. When all traces of medieval times were left behind and the church was not a power into any political power to any further extent, then, a dimly dawn began to appear, to credit for the evidences so copiously collected for centuries.

In modern times, many religions spread all over the world; every single religious group believed, from the very moment of its foundation, they have been in possession of the truth. Then, as a logical corollary, all other belief systems were in darkness. One of the reasons, but not the only one, is the dissimilar interpretation of the Bible. It is read and fingered by millions of people in the world, but every single theistic group approaches different interpretations from the manuscript. Every single interpretation creates a different religious group. This is the way proliferation of belief systems, which use the Bible as guidance, probably occurred. Are any of those groups in factual possession of the divine truth, or the universal truth? It is not for me to say.

Perhaps, the best approach is: the truth is spread all over, in tiny quantities, among all religious groups of the world, each having its small share of the universal truth. Probably, none has the absolute truth, either in their hands or in the best approach of all. Besides, certain evidences are placed in the religious scenario, but they do not pertain to that place. It is a misconception that anything conventional science is unable to explain, automatically lean to theistic domain. One thing is just to believe and another very different is the completeness of individual experiences, which point to another reality, out of the reach of conventional science and difficult to explain based on the accepted experimental procedures; mainly experimental reproducibility of the phenomenon, to some extent.

The question has remained without an answer, though evidences accumulated all over the world despite creed, cultural background, and social status. Are we just biological beings? Does everything from us disappear at death? Is it just a cycle repeated over and over again through human history? It is, to transit as human bodies and after death nothing is left? In 1975, a book written by Dr. Raymond Moody was of a tremendous impact and shed light, with undeniable evidences, on the already known phenomenon, that he named "near death experience" (NDE). Other books by the same author followed (see Recommended Bibliography). A detailed description of the methods he used opened the door to experimental possibilities. Besides, he also included an overwhelming recount of many case study interviews, which were a first hand description of experiences from people who were clinically death, or at imminent death.

Probably, from that year on a true scientific interest began to grow. Different specialists started their own research, with detailed and careful procedures. Recently, Dr. Jeffrey Long has published the results of his investigations that include samples from around the world, in the form of personal descriptions of the NDE. He also detailed the methods he used in his research work. Nowadays, a compelling amount of scientific information on NDE is available. The results obtained are of a tremendous consistency, since the elements that comprise this phenomenon (NDE) are repeated regularly, despite cultural backgrounds, sex, belief system, religious attachments, social status, professional commitments, and age. The principle of repeatability, in all the cases already studied, demonstrate that NDE is a fact. At present, hundreds of cases have been scientifically documented, and the results are astonishingly consistent with a sustain pattern in a high percentage, leaving no doubt of the authenticity of the experiences.

The events or elements present during an NDE, according to most specialists based on the copious case studies, are usually but not limited to an out-of-the-body experience, entering into a tunnel at the end of which appears a light, this light is rapidly enhanced on a mystical essence, the being of light presents to the near-death experiencer his/her life review in a panoramic display, a feeling of an immense love all around, encounters with deceased friends and relatives, an impression of an undetermined time-space reality, a heavenly realm surrounding the person in which colors and lights are of an unearthly nature and then, the return to his/her physical body.

Presently, no academic or official definition has been offered to NDEs events. Nevertheless, any attempt to do so is somehow difficult, since some subtle differences may occur from one person to another during these extra-corporeal experiences. It is not easy to create an integrated-definition that may cover all the details present in all particular experiences up to present. It is my opinion that the evidences, the utterly strong evidences on the phenomenon, are much more important than an academic definition.

All the results already accumulated indicate that science is at an urging process to reconsider many of the "scientific dogmas" regarding the concept of death. Furthermore, probably it shall be advisable to reconsider certain aspects of evolution.

These concepts have been measured, only and exclusively, on the material remains unearthed during decades and through DNA sequencing. But the knowledge obtained from these results is not sufficient to clarify the special material-energy-essence we may represent.

All the exposures from people that experienced NDE clearly point to the fact that consciousness persevere after bodily death. It appears that we are not just a biological mechanism that comes to life, aged through years, and finally die to completely disappear. Consciousness survives physical death. Material life seems to be just a transient state of energy, but it is not exclusively attached to biological, physical and chemical laws; there is another unknown energy that, for the moment time, we can only call *consciousness*, which is beyond the traditional concepts and precepts of official scientific knowledge. Ultimately, you can call that energy soul, spirit, *alter ego* or whichever other appellative you may consider appropriate, what matter is the reality of the filed facts as an incontrovertible truth. The point is that at present, science cannot identify what kind of energy is responsible for the consciousness to exist after physical death. Every aspect reported by near-death experiencers contribute to testify for a very particular kind of energy. According to their descriptions, the colors, brightness, the enhanced perception, and the awareness on the surroundings they perceived are not the same, in a high percentage of study cases, as those we see and perceive in our daily life.

More investigations will come soon, since the results obtained are a real breakthrough in the knowledge of the essence of the human being. From now on, we cannot consider ourselves as highly developed bio-mechanisms evolved and destined to decay on planet Earth. We are an almost perfect, with certain drawbacks, biological model, but not exclusively that. There is a condition of existence, an unknown state of energy beyond the physical barriers and not well understood, which is already evident through all the results from the research work accomplished by well-known scholars up to present. Furthermore, the aftereffects observed due to NDEs produce changes in the behavior of experiencers, and in the way they value life itself from that moment on. They turn to be different; regarding the importance they confer to social status, material belongings. They tend to be more compassionate toward others, besides a completely different approach to the concept of death. It is scholars' opinion that, since the personality of experiencers change, a complete gene test is strongly recommended. Certain studies suggest that genetic expressions may change after certain NDEs. This possibility demonstrates how profound the aftereffects might be.

Scholars are coincident in the criterion that every NDE is unique; regarding its quality, depth, and circumstances. Though general elements are present certain variations are also found. In my very own opinion, one of the most important aftereffects is the "no fear of death" which lasts after a NDE, impregnated in the experiencers for the rest of their three-dimensional existence. One of the most threatening events in humans' life is death.

Of course, many questions shall never be answered. One of those possible questions could be: did any of the *Homo* species before *Homo sapiens,* as *Homo habilis, Homo ergaster, Homo rudolfensis, Homo erectus, Homo heidelbergensis, Homo neanderthalensis,* and *Homo floresiensis* ever had an NDE? We shall never know for certain, fossils are useless to answer such a question, not to mention our complete ignorance as how much evolved was consciousness in those primitive or ancient human forms. Is NDE exclusively associated to a high-developed human condition? This question, *per se,* entails another question of fundamental approach: which *Homo* species could be considered sufficiently evolved as to bear the possibility of NDE's?

Perhaps we have to put down any kind of differentiation among the diverse *Homo* species and consider them all as true humans, regardless of the evolutionary stage they faced to live along the evolutionary pathway. If there is a special kind of energy that represents what we really are, despite the physical body we carry during our lifetime, it is plausible to accept that any human form, no matter how ancient or poor physically evolved it was, could develop the NDEs as well. It is somehow wrong, as if we disregard the possibility of NDEs to persons with any kind of mild retardation or learning deficiency, just because they do not have the same acute understanding or reasoning as the common people.

Possibly, we could not tell one *Homo* species apart from another regarding NDEs, if it were possible at all. Though science is unable to get deeper into these considerations, probably the essence of human condition and the capacity of NDEs date back long before any predictable evolutionary time; probably along the lengthy journey of our ancestors up to our own time. Not less than 185 years ago several authors, from different parts of the world, wrote on this reality evidencing that more than 90 percent of the world cultures acknowledged this phenomenon. In ancient times Egyptians, Greeks, and Tibetan scholars already knew about these extraordinary events. At present, and particularly from 1970, clinical resuscitation procedures have advanced really fast, bringing back to life people that, decades before, were impossible to resuscitate. These technological advances made possible the interviews with near-death-experiencers, together with scientifically designed research and studies. Again, it was Dr. Moody who, for the first time, coined the expression **near-death experience** (NDE).

At present, thirty-six years of sustained research on NDEs have demonstrated that there is a pattern of elements on these experiences pointing to a common denominator that identify this phenomenon. Among the experiencers there are children five years old or younger, aged persons, and multi-cultural case studies. Besides, all skeptics' theories have been disregarded in front of the evidences. Those theories are untenable when confronted with all the detailed researches and the consistent results obtained by each and every one of the scholars involved in these investigations. Some of these theories or models, proposed to dismiss the possibility of NDEs as a transcendental phenomenon, included psychological and physiological theories, taking into account neurochemical and neuroanatomical models.

Altogether, during this period, not less than 4,000 case studies of near-death experiencers have been documented, together with more than 700 scientific articles published in specialized journals as the Journal of Near-Death Studies, among others.

At present, science is at the entrance of an unprecedented kind of new knowledge that will lead to a deep understanding of our most intimate nature. Well-known scholars have conducted abundant researches, and their results point to the fact that we are not only, or exclusively, biological material bodies. It appears that the essence of existence is not contained in our physical structured-bodies. We represent, as our true nature, a kind of energy unknown to science at present. The physical body is like a jewel box into which that essential energy is contained. It points, as many scholars have claimed, that physical life is just a transient state. Our true beings are not attached to our physical appearance. The energy that represents our true existence does not disappear with bodily death. We are a kind of energy that goes beyond the three-dimensional state. I dare to speculate that, if all present results and evidences are enhanced with future researches, chances are that the ultimate conclusion shall be: we are not really biological beings, but energy-beings that belong to a non-three dimensional state, passing through a transient biological phase, which we call life. During our transient biological occurrence, we experienced all constrains of the three-dimensional world. I suggest the Recommended Bibliography at the end of this chapter to all those readers really interested to go further in the knowledge on this field.

I would like to stress the fact that, in general terms, some scientific articles regarding near-death experiencers who have seen dead relatives during their NDEs, are not always specific about the relationship. The researches do not mention if those dead relatives are the experiencer's parents, spouse, brothers and sisters, or any other kin. This unspecific approach prevents the true identification of those beings during the NDEs. If a careful scientific questionnaire was accomplished, why such an important specific detail failed to appear in some results?

The huge amount of evidences and scientifically conducted researches clearly show the outcomes of such energy through the NDE phenomenon; they are just the tip of the iceberg, a long journey lies ahead. The results at hand are promissory and true encouraging for present and future researchers. The substantiation of all these events (NDEs) was clearly documented since ancient times. The nineteenth and twentieth centuries included, as more recent periods. It was not until 1975 that a real boom occurred, as mentioned before. From that on, science entered a new frame of concepts, being mandatory to reconsider many scientific dogmas, which were elaborated without the present knowledge, now involving medical doctors, biologists, and anthropologists among other specialists. In fact, and using just few words, the scientific community keeps itself in a stand-by position regarding NDE. A scientific explanation has not been found to really clarify this phenomenon, at least at the light of orthodox scientific paradigm. This, regardless of the huge amount of cases and evidences already published for more than one third of a century. Probably, but not exclusively, the main obstacle with the greater part of scientific explanations failures, has been its dependence on material explanations and approaches to an occurrence that pertain to a spiritual essence. That energy, that science is unable to grasp the meaning of. Techniques and procedures adequate to explain physical world events are almost useless to approach NDE episodes.

At present, the vast amount of evidences point to the fact that there is no reason to fear bodily death; it is not an end, just the transition to another dimension or energetic state, in our own universe. Again, in front of so many case studies and evidences that have been obtained during a long time, a plausible inference is that science is, in fact, entering a new field of knowledge. It was probably neglected due to the inflexible dogmas of contemporary scientific methods. When Darwin stated that the origin of humans must be found in Africa, his inference was based in the evidence of existing great apes in that continent. Facing that evidence, modern humans would be somehow related in the past with similar, if not identical primitive forms, and no doubt about their existence. Many people, including scientists and priests, attacked Darwin in all possible ways, up to the point that he was almost stigmatized for such a filthy comparison. Most readers derided; scientific community and most of the people were not at the essential level of understanding to admit our humble origin. Anthropocentrism was the supreme idea, any other approach was censured. When a great amount of fossils were collected from different parts of the world along decades, his assertion became an incontrovertible evidence of our primitive past. Such assumptions were based on evidences, not on scientific DNA sequencing at that time, which now represents a customary laboratory procedure.

The huge amount of evidences on the elements that constitute NDEs is more than enough to look at those phenomenological accounts in a very serious way. Furthermore, as has been previously discussed, all skeptical proposals have been dismissed and wiped out, when carefully analyzed by scholars. There is no elbowroom for such skeptical theories. Every single attempt to give explanations on NDEs based on physiological, mental, cultural, psychological, and social and/or cultural grounds, have been discarded as impossible. Historically, always a certain amount of people, scientists among them, have shown themselves reluctant to accept new approaches that are not endorsed by the establishment, no matter how strong are the evidences already at hand.

All descriptions collected and seriously studied from near death experiencers indicate an energy phenomenon at death. This energy is capable to separate from the physical body as the essence that represents the core identity of the being. Important approaches have been developed, giving some indication that a real understanding of quantum physics may be used as a first approach to explain the NDE phenomenon. Furthermore, from the Theory of Relativity it is accepted that matter and energy are transposable, hence, all the evidences on NDEs point to the fact that the essence, which we may call spirit, could be some kind of matter-energy complex form.

Comparisons of different belief systems and elements of the NDE, including aftereffects, are harmonious and equivalent. Just to mention some comparisons scholars have made, it is worth to mention that the following belief systems are coincident with elements of the NDE in a very straight implication: the Baha'i Faith, Islam, Christians, Judaism, Buddhism, and Hinduism, just to mention the most important and relevant.

In addition to all the astonishing evidences already accumulated and supporting NDEs, other investigations have been conducted to explore and induce the so-called After-Death Communications (ADC) with successful results, mainly used to resolve grieving conditions on patients who have lost close friends or family members. It has been proved that the contents in the experiences during an ADC are very similar if not coincident with NDE reported elements. Altogether, these results clearly point to the undeniable existence of an unknown dimension or dimensions where energies, of an unfamiliar nature, are disclosed to the human knowledge.

It seems to be evident that the universe itself is a purposed entity, there is no coincidence or random events that could explain the tremendous and incredible coordination of myriads of events to bring life to Earth, we included. Perhaps, the superb condition of human beings is not exactly or exclusively our unbelievable evolutionary history in the three-dimensional world we use to live on during our material existence. In fact, it could be an extra energy that endures beyond physical death, which elucidation, comprehension, and onset, goes beyond biology.

Recommended Bibliography

Achenbach, J. 2010. Lost Giants. National Geographic 218 (4): 91- 109.

Barrow, J. D., Tipler, F. J., Wheeler, J. A. 1988. The Anthropic Cosmological Principle. Oxford Univ. Press. First Edition. 736 pp.

Borenstein, S., 2011. Cosmic census finds crowd of planets in our galaxy. Ass. Press. Feb. 19.

Botkin, A. L. 2000. The Induction of After-Death Communications Utilizing Eye-Movement Desensitization and Reprocessing: A New Discovery. Journal of Near-Death Studies 18(3): 181- 209.

Croswell, K. 2010. Heart of the Milky Way. National Geographic 218 (6): 92-99.

Diamond, J. 1993. The Third Chimpanzee. The Evolution and Future of the Human Animal. Harper Perennial. New York. 407 pp.

Geary, D. C., 2004. The Origin of Mind: Evolution of Brain, Cognition, and General Intelligence. Washington DC. American Psychological Association. 459 pp.

Gribbin, J., Rees, M. 1991. Cosmic Coincidences. Black Swan Edit. 320 pp.

Guggenheim, B., Guggenheim, J. 1997. Hello from heaven! A new field of research confirms that life and love are eternal. N. Y. Bantam. 416 pp

Hawking, S., Mlodinow, L. 2010. The Grand Design. Bantam Books. N. Y. 198 pp.

Holden, J. M., Greyson, B., James, D. 2009. The Field of Near-Death Studies: Past, Present, and Future. In: The Handbook of Near-Death Experiences-Thirty Years of Investigation. Praeger Publishers, an Imprint of ABC-CLIO, LLC, California. USA. 316pp

Hoyle, F. 1988. The Intelligent Universe. Book Sales. 256 pp.

Kirkwood, T. 2010. Why Can't We Live Forever? Scientific American 303 (3): 42-49.

Long, J. (M.D.), Perry, P. 2011. Evidence of the Afterlife-The Science of Near-Death Experiences. Harper One N. Y. Harper Collins Pub. 215 pp.

Lundahl, C. R., Gibson, A. S. 2000. Near-Death Studies and Modern Physics. Journal of Near-Death Studies 18 (3): 143-179.

Masumian, F. 2009. World Religions and Near-Death Experiences. In: The Handbook of Near-Death Experiences-Thirty Years of Investigation. Praeger Publishers, and Imprint of ABC-CLIO, LLC, California. USA 316pp.

Meacher, M. 2010. Destination of the Species-The Riddle of Human Existence. O Books. John Hunt Publishing Ltd. UK. 244 pp.

Moody, R. A. Jr. (M. D.) 2001(new edit.) Life After Life. Harper One N.Y. Harper Collins Pub.175 pp.

Moody, R. A. Jr. (M. D.) 1978. Reflections on Life After Life. Bantam Books. N. Y. 148 pp.

Moody, R. A. Jr. (M.D.), Perry, P. 1993. Reunions-Visionary Encounters with Departed Loved Ones. Ivy Books. Random House Publishing Groups. 172 pp.

Moody, R. A. Jr. (M.D.), Perry, P. 2010. Glimpses of Eternity. Guideposts. New York. 183 pp.

Morris, D. 1984. The Naked Ape. Dell. N.Y. 205 pp.

Noyes, R. Jr., Fenwich, P., Holden, J. M., Christian, S. R. 2009. Aftereffects of Pleasurable Western Adult Near-Death Experiences. In: The Handbook of Near-Death Experiences-Thirty Years of Investigation. Praeger Publishers, an Imprint of ABC-CLIO, LLC, California. USA 316 pp

Pritchard, J. K. 2010. How We Are Evolving? Scientific American. 303 (4): 41-47.

Relethford, J. H. 2006. The Human Species. An Introduction to Biological Anthropology. Mc Graw Hill. Sixth Edition. 473 pp.

Roth, G., Dicke, U. 2005. Evolution of the Brain and Intelligence. Trends in Cognitive Sciences. 9: 250-257.

Sober, E., Wilson, D. S. 1999. Unto Others: The Evolution and Psychology of Unselfish Behavior. Cambridge, Mass., Harvard University Press. 416pp

Wilson, D. S., Wilson, E. O. 2007. Evolution "for the good of the group". American Scientist. 96: 380-389.

Waters, M. R., Stafford Jr. T. W. 2007. Redefining the Age of Clovis: Implications for the Peopling of the Americas. Science (315) 5815: 1122-1126

Wooten-Green, R. 2002. When the Dying Speak-how to listen to and learn from those facing death. Loyola Press, a Jesuit Ministry. 197 pp.

Epilogue

- It appears that our DNA is not completed enough as to guarantee a more healthy biological, intellectual, and cultural existence. In general terms, it might be said that is an unfinished molecule. Under certain lab conditions multi-stranded DNA chains have been obtained, not precisely related to Watson and Crick model.

- Evolution, in a straight future line, seems to be based on technology, though technology is just an aspect of culture. The possibility to create sperm is already a fact, or at least it is at the corner of the street.

- Apes, on the average, do not suffer from abnormal health conditions as much as humans do. Malformations at birth are not really common. Besides, delivery is an easy physiological process in apes. Human females' pelvis is well designed to upright walking, which is detrimental for a smooth delivery in many cases.

- Our lifespan is now improved based only and exclusively on medical advancements. On the average, our lifespan mean value would be less than that of Neanderthals otherwise, albeit they were exposed to accidental death due to their harsh living and subsistence conditions in the wild. Furthermore, I wonder if the calculations to obtain the mean value of present lifespan also include the huge figures that represent children prematurely dying of cancer, and other cruel health conditions.

- Humans do not seem to be less aggressive as evolution goes by. Nowadays, we are more dangerous than ever before, since we master an enormous technological potential for massive destruction. So huge is the possibility that any of the previous extinctions, known to science, is just a mere dummy run of destruction compare to our destructive potential.

- There are many inferred lineages of descendants in the "accepted" and provisional evolutionary tree leading to *Homo sapiens.* Several questions are still with no apparent answers. Is it a lack of fossils? Is it uncertainty just because real gaps (genetic and paleoanthropological) do exist? What the conclusion shall be if no more important fossils are found? How the answer might be for what is yet unknown? The uncertainty about who were our true ancestors, since we only have a provisional evolutionary tree, proves the insufficient understanding of the biological sequence in our evolutionary history. If we exclusively based this knowledge on fossils, is like supporting scientific understanding on a stochastic process, like "hide and seek." Who knows what the next fossil will prove, if any?

- If the "out of Africa" event would not happen at all, chances are that *Homo sapiens* would never be. Were Neanderthals in the right evolutionary pathway to become as humans as we are today? Or, they were not the chosen ones. Our evolutionary history has been of pain and suffering. Imagine any ancient human species suffering from any painful disease without analgesic of any kind, just in the middle of the wilderness. We may add the perils of predators roaming around. Do you imagine cancer or any other life-threatening painful condition, debilitating those primitive humans without any treatment, suffering with no relief? Why so much pain and suffering to evolve, not only for humans, for every single species on earth.

- Individuals, not populations, have achieved the impulse of science and technology. Populations benefit from inventions and top advancements, though in general terms they are not the producers.

- Though evolution happens along big number of years, about millions and beyond, it is difficult to explain the beauty of nature when we look at flowers, butterflies and many other insects, birds, and coral reef fish. Millions of years of evolution may explain diversification of species, but it is very difficult to explain the existence of beautiful biological products invoking genetic variation, natural selection and evolution processes. Natural selection is not a "sensible artist" to create such marvelous jewels of living creatures. Beauty is not an attribute for fish or birds to mate. They are not endowed with abstract feelings and perceptions as to identify beauty or the need to mate based on beauty; they are not humans. Besides, the attraction for mating in animal species could be based in many other features and behavioral patterns, without the occurrence of beauty.

- A stochastic process of selection, as natural selection is, may diversify to a great length the whole biosphere, but beauty is beyond the possibilities of this process alone. What is in fact the relation between beauty and evolution? Then again, is it just because being so a beautiful bird, or butterfly, or a coral reef fish attracts mates and then beauty was positively selected? How can a butterfly or a fish evaluate beauty? Beauty may have sense, through a natural selection stochastic process, to a handful of species, but not to a huge diversity as it exists on earth. The perception and evaluation of beauty is supposed to be a high-developed property of high-evolved humans. Beyond diversification and adaptation of species, it is dubious that natural selection could mediate to create beauty in such an abundant manner. Our planet is full of beautiful living creatures, let alone the astonishing landscapes all over the world. Beauty is not exactly a product of natural selection, since this process is randomly achieved without sensible feelings toward the essence of beauty. There are many masterpieces of art, in animal and plant kingdoms. Besides, evolution is just the product of a rigorous natural selection process, and nothing else.

- Near-Death experiences (NDE) account for a kind of energy-existence that survives the bodily death, indicating that, at least humans are not a mere biological complex condemned to live a measly lapse on Earth. Researches on After-Death Communication (ADC) prove that this experience is quite similar to NDE, giving a holistic significance and an integrated approach to the concept of another dimension beyond the physical three-dimensional world we use to live in. All evidences indicate that an unknown state of matter-energy is the essence, spirit or soul, that human's cultures have claimed to exist long before written history was accomplished.

- If, by any chance, we accept creation regarding the presence of humans on Earth, that supreme act of life was not only directed to create humans as we are, all of a sudden. It was much more complex than Adam and Eve in the middle of Eden. As long as we know, human forms trudged painfully along the evolutionary pathway to finally step on the stage as modern humans, with a high-developed brain and creativity. Nevertheless, we are still away of being integral humans, our behavior is the culprit.

www.ingramcontent.com/pod-product-compliance
Lightning Source LLC
Chambersburg PA
CBHW071359170526
45165CB00001B/117